ENVIRONMENTAL RESEARCH ADVANCES

WATER CONSERVATION

PRACTICES, CHALLENGES AND FUTURE IMPLICATIONS

ENVIRONMENTAL RESEARCH ADVANCES

Additional books in this series can be found on Nova's website
under the Series tab.

Additional e-books in this series can be found on Nova's website
under the e-book tab.

WATER RESOURCE PLANNING, DEVELOPMENT AND MANAGEMENT

Additional books in this series can be found on Nova's website
under the Series tab.

Additional e-books in this series can be found on Nova's website
under the e-book tab.

ENVIRONMENTAL RESEARCH ADVANCES

WATER CONSERVATION

PRACTICES, CHALLENGES AND FUTURE IMPLICATIONS

MONZUR A. IMTEAZ, M.D.
EDITOR

New York

For permission to use material from this book please contact us:
Telephone 631-231-7269; Fax 631-231-8175
Web Site: http://www.novapublishers.com

Additional color graphics may be available in the e-book version of this book.

Library of Congress Cataloging-in-Publication Data

Water conservation : practices, challenges, and future implications / editor, Monzur A. Imteaz (Faculty of Engineering and Industrial Sciences, Swinburne University of Technology, Melbourne, Australia).
pages cm
Includes bibliographical references and index.
ISBN 978-1-62808-993-6 (hardcover)
1. Water conservation. I. Imteaz, Monzur A., editor of compilation.
TD388.W349 2013
628.1'3--dc23

2013033667

Published by Nova Science Publishers, Inc. † New York

CONTENTS

PREFACE

With climate change and global warming issues, water resource and supply matters are becoming vitally important all over the world. With an ever-increasing global population, the problem is becoming more and more critical. As such, scientists and authorities are calling for increases in water conservation and recycling. This book details some water conservation practices, challenges and future implications. The book chapters are clustered in different groups.

The first three chapters present a detailed mathematical analysis of the expected benefits, outcomes and optimum sizing criteria of rainwater harvesting in different cities with case studies. Chapter 1 highlights the contemporary concept of considering rainfall variability for optimum sizing of rainwater tanks. As a case study, this chapter presents variations in expected water savings due to spatial and climatic variability for the city of Melbourne, Australia using a daily water balance model. This chapter recommends considering climatic variability for these sorts of analyses. Also, for a large city, spatial variability needs to be considered. Chapter 2 presents a detailed economic analysis of the financial feasibility of rainwater tanks, with case studies for different regions of Sydney, Australia. This chapter explains that a low water price in Australia and obscure financial benefits are the main bottlenecks in the wide adoption of rainwater tanks by households connected to a municipal water supply. This chapter recommends that the government rebate/incentive for rainwater tank installation should be increased to make rainwater tanks financially viable to house owners. It also recommends that for a typical detached house in the Sydney urban area, a 5 kL tank is more suitable than the currently recommended 3 kL tank. Also, for peri-urban regions where a municipal water supply is not available, a 5 kL tank will be enough to meet the need for toilet, laundry and drinking water demands about 95% of the time. Chapter 3 examines factors influencing rainwater tank performance. It presents the reliability of rainwater tanks for the daily non-potable water supply for two urban areas of Iran. The reliability of the system is investigated in terms of roof area, tank size, and water demand. The chapter concludes that if the volume of tanks and daily demands of residents are calculated according to the physical condition of buildings and hydrologic condition of the region, rainwater storage in tanks will increase, making it possible to supply more water.

Chapter 4 outlines a multi-criteria decision-making approach for the adaptation of stormwater infrastructure based on future climate related risks. The chapter presents surveying results collected from da variety of stakeholders from three different cities in South East Queensland,Australia. From the survey results, it is observed that priority values for adaptation alternatives varies significantly depending on whether these are based on the combined judgements of all participants or the judgements of individual stakeholder groups. It also recommends theadoption of a hierarchy model that restricts the influeuce of individual stakeholder groups in determining priorities. Chapter 5 provides a thorough review of public acceptance of alternative water sources in Brazil among low-income households. Through interviews and measurement, the chapter presents water end-uses and the water consumption patterns for different plumbing fixtures and appliances. Detailed statistical correlation results are presented for the acceptance data of rainwater and grey water usage according to the number of occupants in each house, the total and per capita income, age, and formal education levels using the Pearson's correlation.

Chapter 6 presents a detailed analysis of the potential challenges of reusing greywater for domestic usage. This chapter describes different types of greywater systems, from the most basic to the most complex, and the integration of rainwater harvesting in greywater reuse systems. Examples of greywater systems that are implemented in several countries are also presented, including information on installation costs and water savings. The chapter includes survey results from a number of different countries, describing public acceptance of greywater reuse, together with the main public concerns and attitudes.

Chapter 7 presents investigation results on the potential changes of soil hydrologic parameters following irrigation with laundry greywater in two types of soils. The soil parameters investigated include hydraulic conductivity, pH, electrical conductivity (EC), porosity, bulk density and capillary pressure-saturation relationship. The experiments were conducted under unsaturated conditions for different synthetic greywater concentrations. It was revealed that soil hydraulic conductivity steadily increases with the greywater concentration for both soils. It was also found that capillary rise decreases with greywater concentration due to the reduction of surface tension.

Chapter 8 outlines the impacts of non-structural measures - water pricing and demand management - in reducing water consumption, presenting case studies in Australia. This chapter examines how water price and demand restriction can assist water conservation in potable urban water supply. It finds that water restrictions have played an important role in Australia as a drought response option in reducing water consumption when reservoir water levels are low. Such restrictions can reduce water demand during the restriction period by about 33% in Australia.

Chapter 9 presents water quality issues of a large water body due to land-use changes within the drainage basin, providing a case study for Lake Hardibo, Ethiopia. It quantifies the spatio-temporal changes in land use and land cover in the closed drainage basin of Lake Hardibo over the last 50 years using multi-temporal remote sensing geospatial data. The chapter concludes by discussing significant consequences for hydrological, soil erosion and sediment processes in the lake's drainage basin system.

Finally, chapter 10 provides in-depth discussions on challenges in the transition toward adaptive water governance. Providing a case study for Cache River Watershed in southern Illinois this chapter discusses various metaphysical, epistemological, institutional, and planning challenges that constrain the transition toward adaptive governance of water resources. The chapter also identifies several key concerns in water resources management such as failure to recognize complexity, neglect of local knowledge, lack of community participation, and a narrow focus of resource management goals.

ACKNOWLEDGMENTS

The editor would like to express appreciation to all those who have contributed to this book through writing different chapters. Special thanks to Ms Miranda Beale, Administration Officer, Centre for Sustainable Infrastructure at the Swinburne University of Technology for her editorial support. Also, the editor expresses gratefulness to Carra Feagaiga, Nadya Gotsiridze-Columbus and others at Nova Science Publishers Inc., NY, USA for their continuous support during the preparation and publication of the book. Gratitude to the editor's son, Miqdaad Imteaz for his help on several editing matters.

About the Editor

Dr. Monzur Imteaz is a Senior Lecturer and Postgraduate Program Coordinator within Civil Engineering group of Swinburne University of Technology at Melbourne, Australia. He has received his B.Sc. in Civil Engineering from Bangladesh University of Engineering & Technology and M.Eng. in Water Resources Engineering from Asian Institute of Technology (Thailand). He has completed his Ph.D. in 1997 on Lake Water Quality Modelling from Saitama University (Japan). After his PhD, he was working with Institute of Water Modelling (Bangladesh) in collaboration with Danish Hydraulic Institute (DHI). Later he has completed his post-doctoral research at University of Queensland, Brisbane. Before joining at Swinburne he has been involved with several Australian local and state government departments.

At Swinburne, Dr Imteaz is teaching subjects 'Urban Water Resources' and 'Integrated Water Design'. Also, he has been actively involved with various researches on sustainability, water recycling and modelling, developing decision support tools, rainfall forecasting using Artificial Neural Networks. He has been serving as a member of the editorial board for several international journals.

In: Water Conservation
Editor: Monzur A. Imteaz
ISBN: 978-1-62808-993-6

Chapter 1

ANALYSIS OF STORMWATER HARVESTING POTENTIAL: A SHIFT IN PARADIGM IS NECESSARY

Monzur Alam Imteaz[1*], Amimul Ahsan[2] and A. H. M. Faisal Anwar[3]

[1]Faculty of Engineering and Industrial Sciences,
Swinburne University of Technology, Melbourne, Australia;
[2]Department of Civil Engineering, Green Engineering
& Sustainable Technology Lab, Institute of Advanced Technology,
University Putra Malaysia, Selangor, Malaysia;
[3]Department of Civil Engineering, Curtin University of Technology, Perth, Australia

ABSTRACT

With increasing population and changing climate regime, water supply systems in many cities of the world are under stress. Water demand is increasing day by day but resources of fresh water are limited. To tackle the situation many water authorities around the world have been promoting the use of water conservation and recycling options through various campaigns and offering incentives/grants for such water saving ideas and innovations. Even with several educational and awareness campaigns and financial incentives, there is a general reluctance to adopt any potential stormwater harvesting measure. The main reasons behind this are that people are not aware of the payback period for their initial investment and the optimum size of the storage required satisfying their performance requirements. Among all the alternative water sources, stormwater harvesting perhaps has received the most attention. One of several water conserving techniques is on-site stormwater harvesting for non-drinking purposes. However there is a lack of knowledge on the actual cost-effectiveness and performance optimisation of any stormwater harvesting system, in particular the proposed design storage volume could be overestimated or underestimated. At present stormwater harvesting systems are proposed and installed without any in-depth analysis of its effectiveness in various climate conditions. The biggest limitation of stormwater harvesting schemes and designs is the

* Faculty of Engineering and Industrial Sciences, Hawthorn, Melbourne, VIC3122, Australia, Email: mimteaz@swin.edu.au.

rainfall variability, which will control the size of the storage needed. Furthermore, with the impacts of global warming and potential climate change, climate variability is expected to increase more. The traditional practice of rainwater harvesting volume/size design is based on historic annual average rainfall data. However, design of rainwater harvesting volume based on annual average rainfall data is not realistic. As a stormwater harvesting system designed considering average rainfall will not provide much benefit for a critical dry period. An in-depth analysis considering different climate regimes (dry, average and wet years) is necessary. A user-friendly tool, eTank was developed to make end-users' decision making process easy, effective and knowledgeable. This chapter presents several case studies within Melbourne (Australia) using eTank for the purpose of rainwater tank optimisation. Outcomes of the case studies are presented in the form of cumulative rainwater saved under different climatic conditions (dry, average and wet years).

Keywords: Rainwater tank, climate change, climate variability, eTank

INTRODUCTION

Among all the stormwater harvesting options, rainwater tanks have been widely studied. Fewkes (1999) conducted studies on residential rainwater tanks in the United Kingdom, producing a series of dimensionless design curves which allows estimation of the rainwater tank size required to obtain a desired performance measure given the roof area and water demand patterns. Vaes and Berlamont (2001) developed a model to determine the effectiveness of rainwater tanks and stormwater runoff using long term area historical rainfall data. Coombes and Kuczera (2003) found that for an individual building with a 150 m^2 roof area and 1-5kL tank in Sydney can yield 10-58% mains water savings (depending on the number of people using the building). According to Coombes and Kuczera (2003), depending on roof area and number of occupants, rainwater tank use can result in mains water annual savings of 18-55kL for 1kL sized tanks and 25-144kL for 10kL sized tanks. In Sweden, Villarreal and Dixon (2005) investigated water savings potential of stormwater harvesting systems from roof areas. Villarreal and Dixon (2005) discovered that a mains water saving of 30% can be achieved using a 40m3 sized tank (toilet and washing machine use only). Ghisi et al. (2007 and 2009) investigated the water savings potential from rainwater harvesting systems in Brazil (South America) and found that average potential for potable water savings of 12-79% per year for the cities analysed.

Coombes (2007) conducted studies on the modelling of the rainwater tanks and the opportunities for effective retention storage using the PURRS (Probabilistic Urban Rainwater and Wastewater Reuse Simulator) water balance model. Following over a decade of research into the quality of rainwater collected from roofs, Coombes (2007) has identified the potential for rainwater to be utilised far more extensively than many government regulators are recommending. Rahman et al. (2012) analysed average water savings, reliability and economic benefits of water savings in Sydney and they have found that the average annual water savings from rainwater tanks are strongly correlated with average annual rainfall. They also outlined that the benefit-cost ratios for the rainwater tanks are smaller than 1.0 without government rebate currently offered in some Australian cities. Muthukumarran et al. (2011) found that use of rainwater inside a home in regional Victoria in Australia can save up to 40%

of potable water use. Mun and Han (2012) developed a design and evaluation method for a rainwater harvesting system and concluded that a design based on sensitivity analysis and proper management of a rainwater harvesting system should be emphasized to improve the operation efficiency. Imteaz et al. (2012a) presented rainwater savings potentials and reliability of rainwater tanks in dry years for Southwest Nigeria. Souza and Ghisi (2012) presented detailed outcomes of rainwater harvesting potentials for thirteen different cities around the world. For their analysis, they have calculated average outcomes from long-term simulations using historical rainfall data. Mehrabadi et al. (2013) studied rainwater harvesting potentials and reliability for three different cities in Iran having three distinct climatic characteristics and they have summarised that rainwater tank reliabilities can significantly vary among different cities (within a country) having different climatic conditions.

Despite positive outcome from many studies, there remains a general community reluctance to adopt stormwater harvesting on a wider scale. Part of the reason for this reluctance can be attributed to lack of information about the effectiveness of a stormwater harvesting system and the optimum storage size required to satisfy the performance requirements under the specific site conditions (Imteaz et al., 2011a). A proper in-depth understanding of the effectiveness of any proposed on-site stormwater harvesting system is often lacking. The predicted change in rainfall patterns in Australia as a result of global warming adds further complexity to planning adequate rainwater harvesting schemes. Furthermore, many studies have used mean annual rainfall data or generated rainfall data in modelling rainwater harvesting system. In an area of highly inter-annual rainfall variability, analyses considering long term mean annual rainfall may not be useful.

Jenkins (2007) developed a computer model for the continuous simulations of amount of rainwater stored in the tank, amount of rainwater used, amount overflowed and amount of mainwater used to top up the tank for household rainwater tanks. The model was used for 12 major cities in Australia using daily rainfall data for the simulations of historical average amounts of the above-mentioned variables. Jenkins (2007) concluded that the climate characteristics of the site have a significant influence on the effectiveness of the rainwater tank. The study also showed that the water savings effeciency is a function of the monthly variation in rainfall. Eroksuz and Rahman (2010) investigated the water savings potential of rainwater tanks in multi-storied residential buildings for three cities in eastern Australia. They have concluded that rainwater tank of appropriate size in a multi-storied building can provide significant water savings even in dry years. They have also proposed equations for predicting annual rainwater savings potentials for those cities. Khastagir and Jayasuriya (2010) analysed reliability of rainwater tanks and calculated the reliabililty using a daily water balance model. They have presented contours of optimum tank sizes for surrounding areas of Melbourne, considering the historical daily rainfall, the demand for rainwater, the roof area for a supply reliability of 90%.

In general commercial rainwater tank stakeholders consider the historial annual average rainfall for a particular city/region to work the amount of potential rainfall to be saved. Undoubtedly this concept is wrong and mis-leading. Not only that it does not consider climate variablity, it also ignores amount of rainwater which will be lost as overflows during heavy rainfalls due to tank volume constraint. In reality, even in an averagae year, amount of rainwater which can be saved would be less than the toatl amount diverted from the roof to the tank. All of these analyses were based on historical daily rainfall data, making an average of cumulative historical savings and other variables. Through such analysis of averaged variables/parameters rainwater tank users do not get an actual range of expected outcomes. With the impacts of climate change, such ranges of actual outcomes are expected to be widen further.

This chapter presents application of a daily water balance model for the optimisation of rainwater tank size. The developed model considered daily rainfall, losses due to leakage, spillage and evaporation, roof area, tank volume, rainwater demand, overflow losses and tank top up requirements in case of shortage. Prime rainwater tank outcomes (i.e. water savings) within a large city are presented in relations to tank volume for three different climatic conditions (dry, average and wet years). Imteaz et al. (2011c) presented similar analysis results for large commercial tanks located near central Melbourne. This chapter presents comparison of water savings for domestic rainwater tanks for three other regions (North, South East and South West) of Melbourne.

Daily Water Balance Modelling

Water balance analysis method is simply based on the theory of continuity, i.e. amount of getting into a system would be either equals to the amount of water getting out, or amount of water getting out plus/minus change of storage amount in the system. This method is commonly used in many water resources systems' modelling and analysis. A daily water balance model, eTank was developed (Imteaz, et al., 2011b), which uses daily rainfall amount, contributing catchment (roof) area, losses (due to leakage, spillage and evaporation), storage (tank) volume and water uses. In this model, the prime input value is the daily rainfall amount for three differeent years (dry, average and wet years). The daily generaged runoff volume is calculated from daily rainfall amount by multiplying the rainfall amount with the contributing roof area and deducting the losses. Generated runoff is diverted to the connected available storage tank. Available storage capacity is compared with the accumulated daily runoff. If the accumulated runoff is bigger than the avaialble storage volume, excess water (overflow) is deducted from the accumulated runoff. Amount of water use(s) is deducted from the daily accumulated/stored runoff amount in a daily basis, if sufficient amount of water is available in the storage. In a situation, when sufficient amount of water is not avaialble in the storage, model assumes that the remaining water demand is supplied from the town water supply. The model calculates daily stormwater use, daily water storage in the tank, daily overflow and daily town water use. In addition, model calculates accumulated annual rainwater use, accumulated annual overflow and accumulated annual town water use. The overall process can be mathematically described as follows:

Cumulative water storage equation,

$$V_t = I_t + V_{t-1} - D_t \quad (1)$$

$$V_t = 0, \text{ for } V_t < 0 \quad (2)$$

$$V_t = TV, \text{ for } V_t > TV \quad (3)$$

where,

V_t is the cumulative water stored in the rainwater tank (L) after the end of tth day
I_t is the harvested rainwater (L) on the tth day
V_{t-1} is the storage in the tank (L) at the beginning of tth day
D_t is the daily rainwater demand (L), and
TV is the volume of rainwater tank (L)

Townwater use equation,

$$C_t = D_t - V_t, \; for \; V_t < D_t \quad (4)$$

where, C_t is the townwater use on tth day (L).

Overflow equation,

$$SP_t = S_t - TV, \; for \; S_t > TV \quad (5)$$

where, SP_t is the spillway/overflow on tth day (L).

Reliability is calculated with the equaiton,

$$R_e = D_w/D_d \times 100 \quad (6)$$

where,

R_e is the reliability of the tank to be able to supply intended demand (%)
D_w is the number of days in a year the tank was able to meet the demand, and
D_d is the total number of days in a particular year

EFFECTS OF CLIMATE VARIABILITY

Inter-annual variations of rainfalls magnitudes and patterns significantly influence rainwater tank outcomes. As for example a rainwater tank in a typical dry year will not be able to provide significant benefits, as it will be providing in a wet/average year. With the impact of climate change, this sort of climate variability is getting prominent in many parts of the world. However, most of the published studies assumed many years of rainfall data and eventually produced single averaged outcomes for a particular rainwater tank. With the prominent impacts of climate change, this sort of average analysis will provide misleading information to the users. Users would expect to save same amount of water in every year, irrespective of having climate variability. Imteaz et al. (2013, 2012a, 2012b, 2011a) presented several studies showing variations of rainwater tank reliabilities depending on climatic conditions (i.e. dry, average and wet years) for different regions of Melbourne. However, reliability is more important in rural areas where rainwater tank may be the sole water source in many cases. For major cities like Melbourne, it is more important to analyse potential water savings from rainwater tanks, as in in the major cities there is public water supply system in place, which can be used for augmenting water supply when rainwater tank is empty or not enough to supply intended demand. Figures 1-3 show ranges of expected water savings from rainwater tanks under three different climatic conditions for three different regions (North, South East and South West) of Melbourne. In all the mentioned analyses it was assumed that rainwater tank is connected to a roof area of 200 m^2 and the rainwater demand is 300 L/day. Table 1 summarises the important rainfall statistics for the three study regions, which shows that highest amount of annual rainfall (mean: 862mm) occurs in South-East Melbourne and lowest amount (mean: 543mm) in South-West Melbourne. Mean rainfall in North Melbourne is 621mm. Through statistical analysis of total annual rainfall data, three separate years were selected as dry year, average year and wet year. Selected years and corresponding annual rainfall values are shown in Table 2.

Table 1. Historical rainfall characteristics of selected regions

	North	South-West	South-East
Mean (mm)	621	543	862
Median (mm)	618	539	847
Standard Deviation	130.7	123.3	160
Maximum (mm)	838.7	809	1237
Minimum (mm)	430	295	513

Table 2. Selected rainfalls and corresponding years for the dry, average and wet years

	North		South-West		South-East	
	Rainfall (mm)	Year	Rainfall (mm)	Year	Rainfall (mm)	Year
Dry year (10 %ile)	461	2009	374	1982	588	1967
Avg. year (50 %ile)	608	2001	538	1969	835	1983
Wet year (90 %ile)	729	1977	712	1964	1081	1992

Expected water savings are shown in Figure 1 for North Melbourne, in Figure 2 for South-West Melbourne and in Figure 3 for South-East Melbourne. From all the figures, it is evident that significant variations of water savings are expected within different climatic conditions. Highest variation was observed for South-West Melbourne; expected annual water savings in wet year is almost 1.5 times of expected annual savings in dry year (Figure 2). For South-East Melbourne expected annual water savings in wet year is almost 1.2 times of expected annual savings in dry/average year (Figure 3). Also, for South-East Melbourne, performances of dry year and average year are found to be very similar. This is because of the typical rainfall pattern on those particular years. Sometime, an average year can provide benefits closer to a dry year if the daily rainfall patterns are more sporadic, i.e. some huge burst of rainfalls in several days, which causes rainwater tank to become full and huge overflow losses occur. Because of this reason, just selecting a single year for each of the dry, average and wet years may not be correct representations. Further investigations are underway to have a better representation of dry, average and wet years.

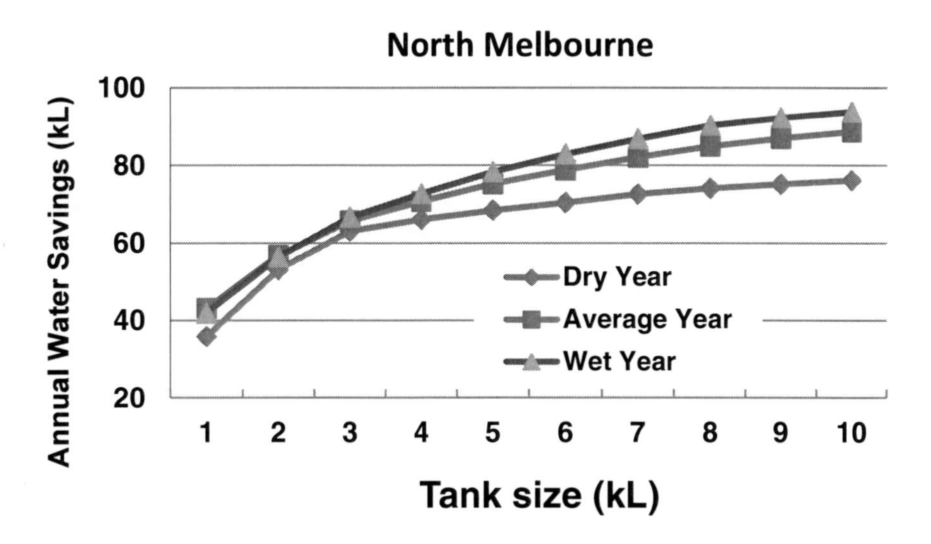

Figure 1. Expected rainwater savings for North Melbourne under different climatic conditions.

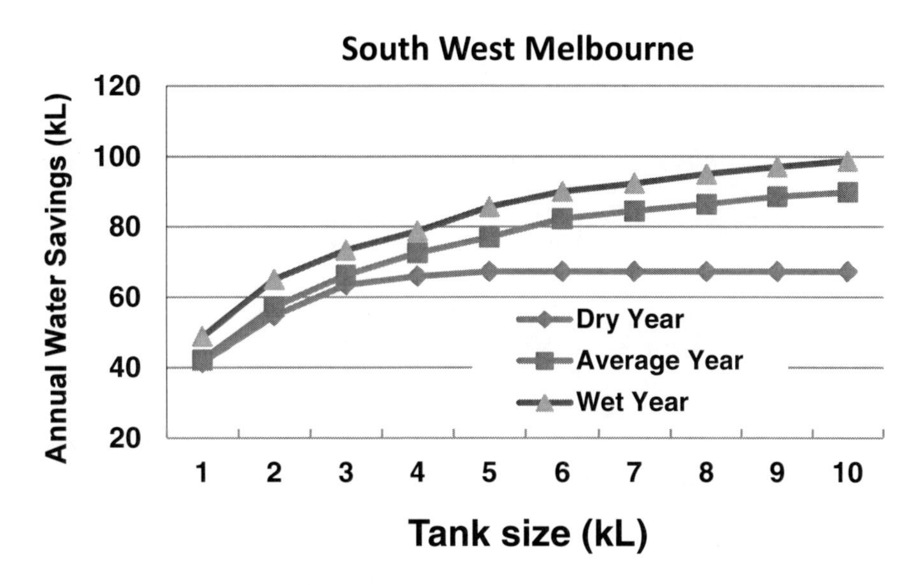

Figure 2. Expected rainwater savings for South-West Melbourne under different climatic conditions.

Effects of Spatial Variability

In this section, spatial variability does not mean variability among different cities; rather it is variability within a city, especially for a large city. Also, it does not mean the variability of rainfalls, rather it mean variability in rainwater harvesting outcomes due to variability of rainfalls and rainfall patterns. Effects of spatial variability in regards to rainwater tank reliabilities were presented by Imteaz et al. (2013) for different regions of Melbourne. In this chapter, effects of spatial variability in regards to expected rainwater savings are presented for different regions of Melbourne. Comparisons of expected water savings are shown in Figure 4 for dry years and in Figure 5 for wet years.

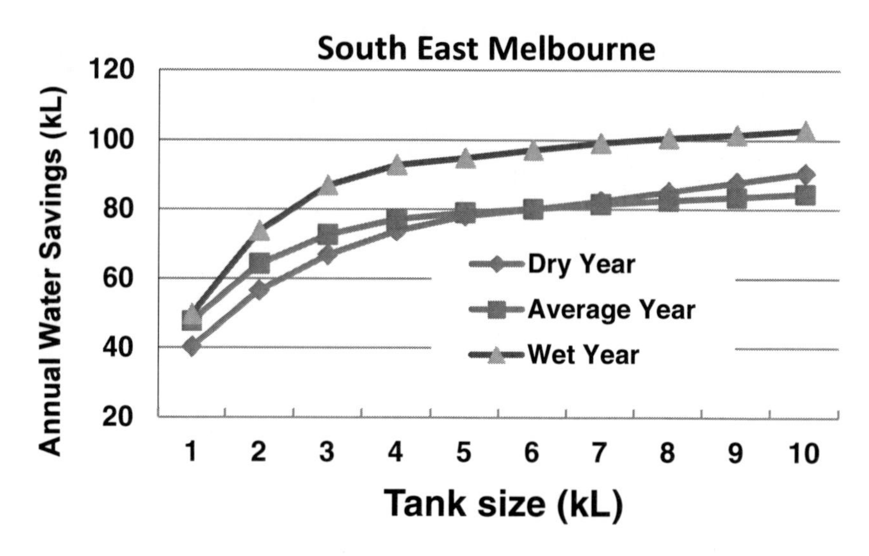

Figure 3. Expected rainwater savings for South-East Melbourne under different climatic conditions.

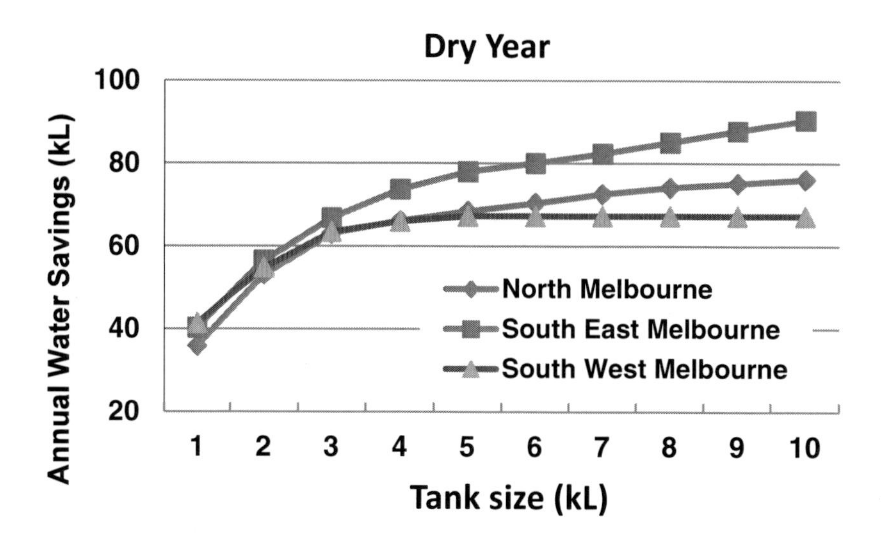

Figure 4. Expected rainwater savings in dry years for different regions of Melbourne.

From Figure 4 it is clear that moderate variations of water savings can be expected within different regions of Melbourne and as expected magnitude of spatial variations are larger for a larger tank size. In a dry year, for a 10kL tank expected water savings for South-East Melbourne is 1.35 times higher than the expected water savings for South-West Melbourne. In reality, these sorts of variations are expected for any large city. However, for a wet year, magnitudes of variations are bit different; for a tanks sizes of upto 4kL, variations increases with the increase of tank size and variations decreases with tank size for tanks larger than 4kL. It is due to the fact that in water year, as rainwater being abundant with larger tank sizes expected water savings will be closer to optimum, unless a higher demand is considered. In a wet year, for a 4kL tank expected water savings for South-East Melbourne is 1.3 times higher than the expected water savings for North Melbourne and for a 10kL tank expected water savings for South-East Melbourne is 1.1 times higher. In regards to expected annual water savings, it may seem that spatial variations are not highly significant. However, for rainwater tanks, 'reliability' is another factor which needs to be considered as well as reliability provides an insight of number of days in a year when an alternate water supply would be necessary for the intended demand. Imteaz et al. (2013) found that reliabilities vary significantly within different regions of Melbourne. It is found that for a standard household (100 m^2 roof having two people), reliabilities vary 22.5% to 73% within the different regions of Melbourne. However, in a higher demand scenario (100 m^2 roof having four people), reliabilities drop (11~18% in an average year) significantly and do not vary significantly within the regions. With a higher rainwater accumulation potential scenario (200 m^2 roof having two people), it is found that across all the regions, in average years it is possible to achieve reliabilities within 80~90% and in wet years it is possible to achieve reliabilities within 90~100% with a tank size ~10000L.

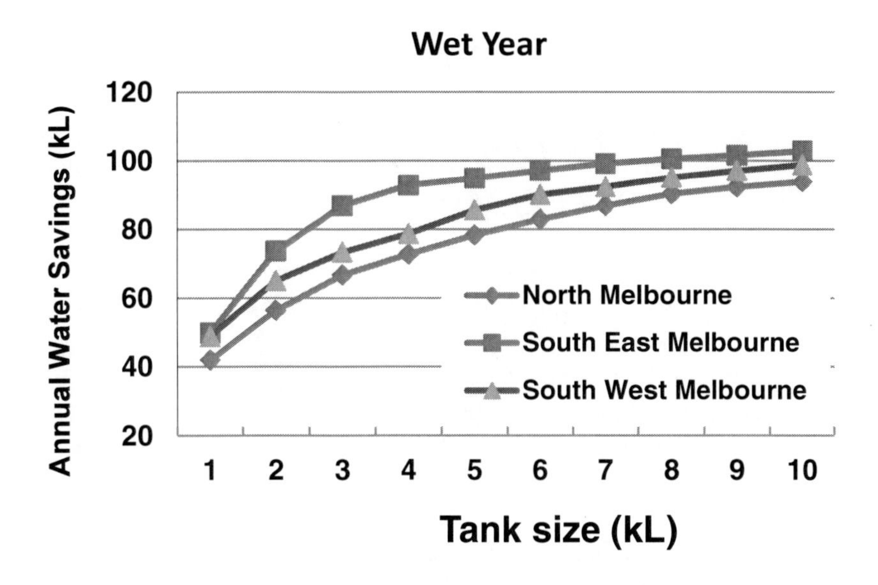

Figure 5. Expected rainwater savings in wet years for different regions of Melbourne.

Conclusion

This chapter investigates expected water savings of rainwater tanks within three different regions of Melbounre, under different climatic conditions. It is found that the results vary significantly among different climatic conditions (i.e. dry, average and wet years). Also, rainwater tank outcomes vary significantly among different regions of Melbourne. These findings emphasize the need to change the traditional design practice of considering a single annual rainfall value for the purpose of rainwater tank sizing. For a small city, climatic variability would be expected. For a large city, in addition to climatic variability, spatial variability also becomes significant. Typical design procedure does not consider these spatial and climatic variabilities, as such it does not always provide an optimum solution. The results presented in this study would vary with geographical locatoins i.e. with different climatic conditions or in general with different rainfall intensities and pattern. In reality, there can be numerous optimal solutions with different combinations of storage volumes, roof sizes, rainwater demand and number of people in the household. A comprehensive decision support tool would assist to achieve the best solution for any geographical location by identifying the optimum storage volume for a given roof area.

It should be noted here that expected water savings is not the only guiding factor in selecting a tank size; factors such as tank space, cost and reliability of intended water supply would also influence the selection of a particular tank size. Imteaz et al. (2013) found that to be able to achieve a suitable reliability, it is recommended to use either a large tank (10000L or more) or connect with a larger roof area (>250 m^2) with a reasonable tank size (>5000L). In general, with the assumed non-potable water demand (150 L/person/day) it is difficult to achieve 100% reliability. In reality, in many parts of the world, the non-potable water demand would be much lower.

REFERENCES

Coombes P. (2007) Rainwater Research, *www.rainwaterharvesting.com.au*; viewed 13.05.10.

Coombes, P., Kuczera, G. (2003) Analysis of the performance of rainwater tanks in Australian capital cities. *28th International Hydrology and Water Resources Symposium*, Wollongong, NSW, Australia.

Eroksuz, E., Rahman, A. (2010) Rainwater tanks in multi-unit buildings: A case study for three Australian cities. *Res., Con. & Rec.* 54, 1449–1452.

Fewkes, A. (1999) The use of rainwater for WC flushing: the field testing of a collection system. *Build. and Env.* 34(6), 765-772.

Ghisi, E., Bressan, D.L., Martini, M. (2007) Rainwater tank capacity and potential for potable water savings by using rainwater in the residential sector of southeastern Brazil. *Build. and Env.* 42, 1654-1666.

Ghisi, E., Tavares, D.F., Rocha, V.L. (2009) Rainwater harvesting in petrol stations in Brasilia: Potential for potable water savings and investment feasibility analysis. *Res., Con. & Rec.* 54, 79-85.

Imteaz, M.A., Adeboye, O., Rayburg, S., Shanableh, A. (2012a) Rainwater harvesting potential for southwest Nigeria using daily water balance model. *Res., Con. & Rec.* 62, 51-55.

Imteaz, M.A., Ahsan, A., Naser, J., Rahman, A. (2011a) Reliability Analysis of Rainwater Tanks in Melbourne using Daily Water Balance Model. *Res., Con. & Rec.* 56, 80-86.

Imteaz, M.A., Ahsan, A., Shanableh, A. (2013) Reliability analysis of rainwater tanks using daily water balance model: Variations within a large city. *Res., Con. & Rec.77, 37-43.*

Imteaz, M.A., Rahman, A., Ahsan, A. (2012b) Reliability analysis of rainwater tanks: A comparison between South-East and Central Melbourne. *Res., Con. & Rec.* 66, 1-7.

Imteaz, M.A., Rauf, A., Aziz, A. (2011b) eTank: A decision support tool for optimizing rainwater tank size, *International Congress on Modelling and Simulation MODSIM*, Perth, Australia.

Imteaz, M.A., Shanableh, A., Rahman, A., Ahsan, A. (2011c) Optimisation of Rainwater Tank Design from Large Roofs: A Case Study in Melbourne, Australia. *Res., Con. & Rec.* 55, 1022-1029.

Jenkins GA. (2007) Use of continuous simulation for the selection of an appropriate urban rainwater tank. *Aus. J. Wat. Res.* 11(2), 231–46.

Khastagir, A., Jayasuriya, N. (2010) Optimal sizing of rain water tanks for domestic water conservation. *J. Hyd.* 381, 181-188.

Mehrabadi, M.H.R., Saghafian, B., Fashi, F.H. (2013) Assessment of residential rainwater harvesting efficiency for meeting non-potable water demands in three climate conditions. *Res., Con. & Rec.* 73, 86-93.

Mun, J.S., Han, M.Y. (2012) Design and operational parameters of a rooftop rainwater harvesting system: definition, sensitivity and verification. *J. Env. Mgt.* 93, 147-153.

Muthukumaran, S., Baskaran, K., Sexton, N. (2011) Quantification of potable water savings by residential water conservation and reuse – A case study. *Res., Con. & Rec.* 55, 945-952.

Rahman A., Keane J., Imteaz M. A. (2012) Rainwater harvesting in Greater Sydney: Water savings, reliability and economic benefits. *Res., Con. & Rec.* 61, 16-21.

Santos, C., Taveira-Pinto, F. (2013) Analysis of different criteria to size rainwater storage tanks using detailed methods. *Res., Con. & Rec.* 71, 1-6.

Souza, E.L., Ghisi, E. (2012) Potable Water Savings by Using Rainwater for Non-Potable Uses in Houses. *Water 4*(3), 607-628.

Vaes, G., Berlamont, J. (2001) The effect of rainwater storage tank on design storms. *Urb. Wat.* 3, 303-307.

Villarreal, E.L., Dixon, A. (2005) Analysis of a rainwater collection system for domestic water supply in Ringdansen, Norrkoping, Sweden. *Build. and Env.* 40, 1174-1184.

In: Water Conservation
Editor: Monzur A. Imteaz

ISBN: 978-1-62808-993-6

Chapter 2

RAINWATER HARVESTING AS A MEANS OF WATER CONSERVATION: AN AUSTRALIAN EXPERIENCE

Ataur Rahman, Joseph Keane and Khaled Haddad*
University of Western Sydney, Australia

ABSTRACT

Rainwater tank has emerged as an important water savings element in urban areas of Australia. Due to greater environmental awareness and government rebate, the uptake of rainwater tank in Australia has been increasing. From the literature it is hard to make any conclusion whether the rainwater tank is economically viable to house owners, which is mainly due to low price of urban water supply. It is thus recommended that rebate for rainwater tank should be increased to make rainwater tanks financially viable to house owners. This will help to achieve a sharp rise in the uptake of rainwater tanks. From a case study in Sydney, it has been found that for a typical detached house, a 5 kL tank is more suited than the currently recommended 3 kL tank. Also, where Sydney Water supply is not available, e.g. in peri-urban regions of Sydney (such as rural areas adjacent to Richmond, Penrith, Cambelltown and Hronsby), a 5 kL tank will be enough to meet the needs for toilet, laundry and drinking water demands for about 95% of time.

Keywords: Rainwater tanks, stormwater, water recycling, water sensitive urban design

INTRODUCTION

Rainwater is generally fresh in nature. However, it gets polluted from atmospheric particles (such as acid rain) and pollutants deposited on impervious surfaces. Overall, rainwater is purer than many other alternative forms of water such as grey water and recycled wastewater. Rainwater if collected and stored, it can meet the demand of some non-potable water usages such as irrigation, toilet flushing, and washing of cars and hard surfaces. Many

* School of Computing, Engineering and Mathematics, University of Western Sydney, Building XB, Kingswood, Locked Bag 1797, Penrith, NSW 2751, Australia Email: a.rahman@uws.edu.au.

residences in Australia show reluctance in adopting a rainwater harvesting system (RWHS). Data from the Australian Bureau of Statistics (ABS) show that about 47% say that the main reason for not installing a rainwater tank is the perceived 'higher cost' (ABS, 2007). The financial incentives provided by Government authorities in the form of rebate to the home owners for encouraging them to install rainwater tanks have helped to enhance the uptake of RWHS in Australia. This chapter highlights the benefits of RWHS as a means of water conservation. At the beginning, the introduction section covers the literature review, which is followed by a review of the status of rainwater tanks in Australia. A case study of rainwater tanks in the city of Sydney, Australia is then presented. Finally, a conclusion from the chapter is provided.

There have been many studies showing the benefits that RWHS can provide (Van der Sterren et al., 2012; Rahman et al., 2012a, b; Imteaz et al., 2011a, b; Zhang et al., 2009, 2012). For example, Muthukumarran et al. (2011) found that use of rainwater tank inside a purpose-built home in regional Victoria can save up to 40% of potable water use. Imteaz et al. (2011a) demonstrated that for commercial tanks connected to large roofs in Melbourne, total construction costs can be recovered within 15 to 21 years. Tam et al. (2010) found that RWHS can offer notable financial benefit for Brisbane, the Gold Coast and Sydney due to the relatively higher rainfall in those cities as compared to Melbourne. Zhang et al. (2009) examined the viability of RWHS in high-rise buildings in four capital cities in Australia and found that Sydney has the shortest payback period (about 10 years) followed by Perth, Darwin and Melbourne. Khastagir and Jayasuriya (2011) conducted the financial viability of RWHS in Melbourne and found that payback period vary considerably with the tank size and local rainfall. For multi-storey buildings Eroksuz and Rahman (2010) found that in order to maximise the water savings, a larger tank would be more appropriate and that these tanks could provide significant water savings, even in dry years. Khastagir and Jayasuriya (2009) used water demand and roof area to develop a set of dimensionless number curves to obtain the optimum rainwater tank size for a group of suburbs in Melbourne. Imteaz et al. (2011a, b; 2012) demonstrated the usefulness of large rainwater tanks in Melbourne city.

Domenech and Sauri (2010) carried out the financial viability of the RWHS in single and multi-family buildings in the metropolitan area of Barcelona (Spain) and found that in single-family households an expected payback period would be between 33 to 43 years depending on the tank size. In contrast, for a multi-family building a payback period is as high as 61 years for a 20 m^3 tank. Ghisi et al. (2009) found that an increase in the tank size enhanced the reliability of the rainwater tank notably in meeting the demand for car washing. In another study, Kyoungjun and Chulsang (2009) found that rainwater collection is only feasible in South Korea during six months of the year and benefit cost ratio higher than 20% could not be achieved.

For multi-storey buildings in Sydney, Rahman et al. (2010) found that it could be possible to achieve "payback" for the RWHS under some scenarios. It was found that a smaller discount rate would be more favourable and the greater the number of users the higher the benefit-cost ratio for a RWHS. Ghisi and Schondermark (2013) found that the ideal tank capacity can be conservative for high rainwater demands and in these cases, an investment feasibility analysis should be carried out in order to obtain a more appropriate tank capacity. It was noted that rainwater usage would be economically feasible for most cases; and the higher the rainwater demand, the higher the feasibility. In a study in Iran, Mehrabadi et al. (2013) found that in humid climate, it is possible to supply about 75% of non-potable water demand by storing rainwater from larger roof areas for a maximum duration of 70% of the times. For roofs with small surface area, the supply can meet 75% of non-potable water demand for a maximum duration of 45% of the times. For Mediterranean climate, it is possible to supply at least 75% of non-potable water demand in buildings with larger roof areas for a maximum duration of 40% of the times. However, in arid climate, similar duration is only 23% of the times.

Campisano and Modica (2012) presented a dimensionless methodology for the optimal design of domestic RWHS. The methodology was based on the results of daily water balance simulations carried out for 17 rainfall gauging stations in Sicily, Italy. In this study, a dimensionless parameter to describe the intra-annual rainfall patterns was introduced and easy-to-use regional regressive models were developed to estimate water savings and overflows from RWHS. They found that the economical convenience of large tanks would decrease as rainwater availability decreases.

Zhang et al. (2012) examined how rainwater tanks can be used to mitigate water logging problem in Nanjing, China. The results showed that exploitation of rainwater harvesting from rooftops and other underlying surfaces has high potential. They showed that urban waterlogging problems can be effectively reduced through rainwater harvesting by 13.9%, 30.2% and 57.7% of runoff volume reduction in three cases of the maximum daily rainfall (207.2 mm), the average annual maximum daily rainfall (95.5 mm) and the critical rainfall of rainstorm (50 mm).

STATUS OF RAINWATER TANKS IN AUSTRALIA

About 93% of Australian households have access to mains water as in March 2007 (ABS, 2007). Where mains water is not available, people source their water from rainwater tanks, groundwater wells or other sources as shown in Figure 1. Due to frequent droughts and greater environmental awareness, the majority of Australian households have some sort of water conservation devices. As reported by the ABS, in June 1994, only 39% of households had a dual flush toilet. In contrast, in 2007, 81% of households had a dual flush toilet. The percentage of dwellings using water-efficient shower heads has risen from 22% in 1994 to 55% in 2007. These data show a sharp rise in the use of rainwater tanks in Australia.

Grey water is recycled from the shower/bath, laundry or kitchen water usages. It has been reported by ABS that in 2007, grey water was the second most common source of water for households, after mains/town water. About 54% of Australian households use grey water as a water source. It has been noted that Victoria had the highest percentage of households using grey water as a source (72%), followed by the Australian Capital Territory (63%). The Northern Territory had the lowest reported use of grey water (32% of households). In 2007, about 24% of Australian households use grey water as their primary source of water for the garden. It has been reported that Victoria and Queensland have the highest grey water usages for garden (43% and 27%, respectively). The grey water usage rates in garden for other Australian states are: Australian Capital Territory (21%), New South Wales (19%), the Northern Territory (4%) and Western Australia (5%).

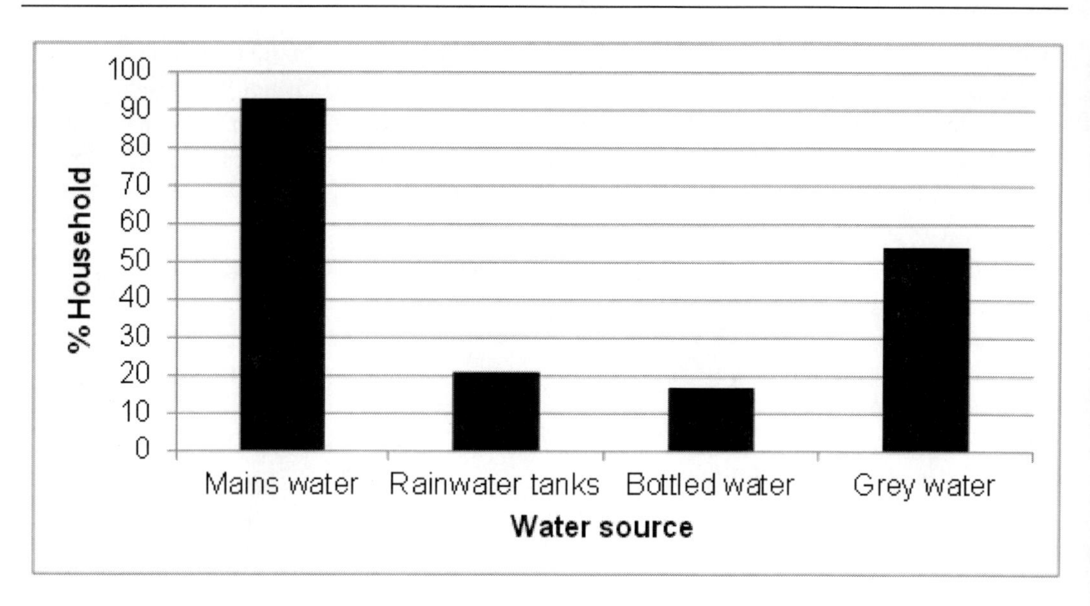

Figure 1. Sources of water for Australian household as per 2007 data (ABS, 2007).

As reported by ABS, in 2007, 21% of all households have a rainwater tank. South Australia has the highest proportion of dwellings with a rainwater tank (49% total), and the Australian Capital Territory and the Northern Territory have the lowest proportions (8% and 6%, respectively). Rainwater tanks are more common outside capital cities (35%) than within capital cities (12%). In capital cities, the most commonly reported reason for installing a tank was to save water; however for rural areas, the most common reason is that the area does not have mains water. About 42% of households with a rainwater tank report saving water as a reason for installing a tank, and 27% reported that their household was not connected to mains water. It has been reported by the ABS that more than 60% of households without a rainwater tank (but which had a dwelling suitable for a tank and which were home owners or purchasers) had considered installing one; however, cost was the most common reason for not having a rainwater tank (48%). The reason for installing a rainwater tank is illustrated in Figure 2.

After 2007 ABS report on rainwater tanks, an update was released by ABS in 2010 (ABS, 2010). According to 2010 data, about 26% of households used a rainwater tank as a source of water in March 2010, up from 21% in 2007. About 49% of South Australian households used a rainwater tank, followed by Queensland (36%) and Victoria (30%). The Northern Territory had the lowest proportion of households (5%) that used a rainwater tank. During the period of 2007 to 2010, the greatest increase in rainwater tank use was in Queensland and Victoria. ABS (2010) stated that of households living in a dwelling suitable for installation of a rainwater tank, the percentage that have a rainwater tank installed has increased from 24% in 2007 to 32% in 2010, i.e. 8% increase in three years.

Brisbane had the largest increase of households that have a rainwater tank at their dwelling, from 18% in 2007 to 43% in 2010. It was also reported that the proportion of households residing at a dwelling less than one year old that have a rainwater tank installed has risen to 57% in 2010, compared with 26% in 2007. The reasons reported by Australian households for installing a rainwater tank were 'to save water' (47%) and 'restrictions on mains water' (24%).

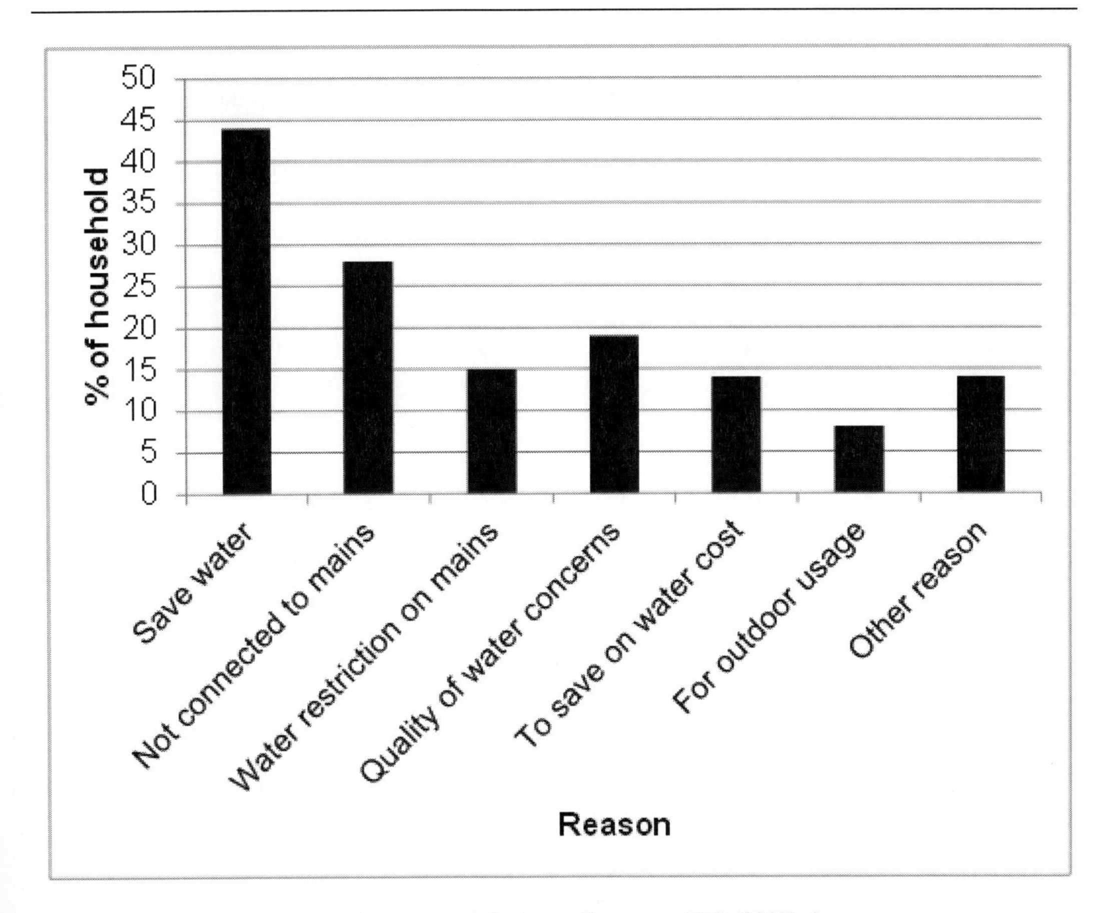

Figure 2. Reason for having a rainwater tank in Australia as per ABS (2007) data.

Melbourne had the highest proportion of households (47%) to report water restrictions as a reason for installing a rainwater tank. These data shows that RWHS is becoming a pupular water conservation and reuse means in Australia except for the Northern Territory.

A Case Study for Sydney, Australia

In this case study, the reliability of various rainwater tank sizes for detached houses in Sydney, Australia is examined. Sydney is the largest city in Australia, with a population of over 4.5 millions. The water supply of this city is largely dependent on surface water reservoirs. RWHS can serve two purposes in Sydney. Firstly, this can reduce the demand of mains water by providing on-source non-potable water such as toilet flushing and irrigation, which cover about 50% of the household water demand. Secondly, RWHS can offer a greater resilience to Sydney water supply system by offering alternative drinking water in case the major reservoir system is contaminated.

In Sydney, as per BASIX requirement, every new house must have a RWHS. BASIX refers to the planning regulation of the New South Wales Government to enfornce water and energy savings measures to home owners. Currently, it is a 3 KL tank that is generally needed for a detached house. Since Sydney is a very large city and has a high rainfall gradient, with a higher rainfall in the coastal area and smaller values in the western area, it is likely that different parts of the City need different tank sizes for achieving the best possible water savings. This case study investigates the water savings potential and reliability of water supply for RWHS in detached houses at different locations in Greater Sydney. This will assist in determining a suitable tank size at a given location in Sydney.

Ten different locations in Sydney are considered as shown in Figure 3. These are Bankstown, Campbelltown, Cronulla, Hornsby, Kellyville, Manly, Penrith, Parramatta, Richmond and Sydney City Centre. The daily rainfall data at these locations is obtained from the Australian Bureau of Metrology (BOM). The rainfall data length ranges from 31 to 150 years (average: 73 years). The average monthly rainfall distribution in the study region is presented in Figure 4, which shows the lowest rainfall occur during July and August, which are winter months, when irrigation demand is expected to be at the lowest level.

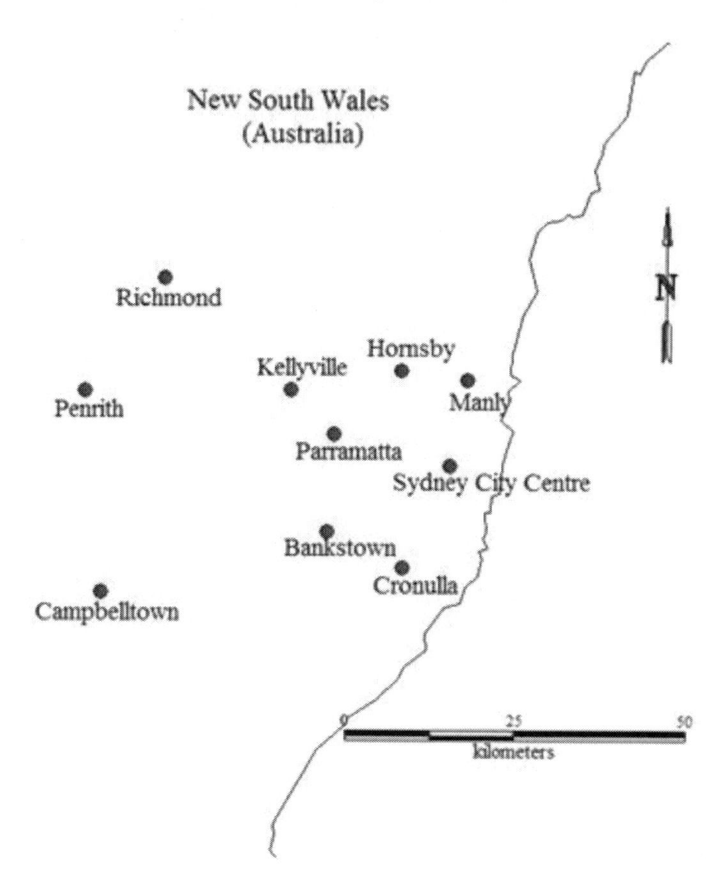

Figure 3. Ten study locations in Sydney (Australia).

In this case study, three different combinations of water usages are considered: (a) toilet and laundry (b) irrigation and (c) a combination of toilet, laundry and irrigation (combined use). Three different tank sizes are considered, which are 2 kL, 3 kL and 5 kL. A hypothesized new development is considered at each of the study locations with a detached single household having 4 occupants. A total site area of 450 m^2 is considered with a roof, lawn and impervious areas of 200 m^2, 150 m^2 and 100 m^2, respectively. It is assumed that the rainwater tank is constructed above ground, with a concrete base.

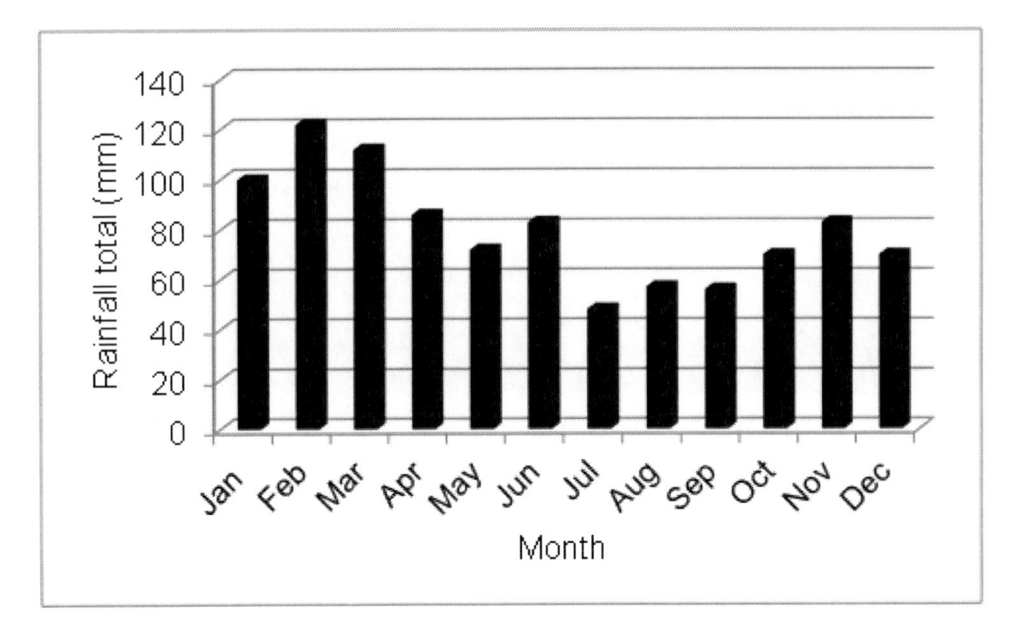

Figure 4. Average monthly rainfalls in Sydney (based on the data by BOM, 2010).

It is also assumed that a mains top up system supplements the tank once the volume of water drops below 20% of the tank capacity. A pump is assumed to be serving indoor (toilet and laundry) use, while a gravity system is applied for irrigation use. To achieve better water quality, first flush and leaf-eater devices are considered to filter contaminants from roof runoff before water enters into the tank as per local standard.

Data for toilet, laundry and irrigation usages for residential houses are obtained from Sydney Water. A 6-litre AAA (the higher the A's the more water efficient the device is) rated dual flush toilet is assumed with a frequency of toilet use of three times per person per day, giving 0.018 kL per person per day water demand. A washing machine, rated 4A, is assumed that has a volume of 50 litres and it is used 3 times every week, which is equivalent to 0.0215 kL/day of water use. An irrigation demand (for lawn and other garden irrigation) of 10 mm per day per square metre of lawn area is also assumed.

To estimate the water savings, a water balance simulation model (WBSM) on a daily time scale in excel was built, which incorporated various elements of RWHS including tank size, daily rainfall, losses, daily water demand, mains top up and tank spillage.

Here, daily rainfall is regarded as inflow and the release (i.e. water supply) as well as possible spill as outflow following the approach by Su et al. (2009). Here release is estimated using the following equations:

$$R_t = D_t \text{ if } I_t + S_{t-1} \geq D_t \tag{1}$$

$$R_t = I_t + S_{t-1} \text{ if } I_t + S_{t-1} < D_t \tag{2}$$

where D_t is the daily demand (m^3) on day t, S_{t-1} is the tank storage at the end of the previous day (m^3), R_t is release from rainwater tank (m^3) and I_t is inflow (m^3). Spill (SP_t) (m^3) is calculated from the following equations:

$$SP_t = I_t + S_{t-1} - D_t - TV \text{ if } I_t + S_{t-1} - D_t > TV \quad (3)$$

$$SP_t = 0 \text{ if } I_t + S_{t-1} - D_t \leq TV \quad (4)$$

where TV is the design storage capacity (m^3) of the tank. The tank storage S_t at the end of day t can be calculated using the following equations:

$$S_t = TV \text{ if } SP_t > 0 \quad (5)$$

$$S_t = S_{t-1} + I_t - R_t \text{ if } SP_t \leq 0 \quad (6)$$

The reliability of a RWHS is calculated as the ratio of the number of days when desired demand is met fully by the available rainwater provided by the RWHS and the total number of simulated days considered in the study.

Water savings data for the three different water usages are presented in Tables 1, 3, 5 and 7. These show that for a given tank size water savings for toilet and laundry usages exhibit little variation across different locations in Sydney. However, there are notable differences in water savings for irrigation usage and 'toilet, laundry and irrigation' usage. As shown in Table 7, Manly has the highest water savings and Richmond has the lowest one where Manly exhibits 28%, 35% and 48% higher water savings than that of Richmond for 2 kL, 3 kL and 5 kL tank sizes. It is noted that Manly and Richmond have the highest and lowest average annual rainfall values of 1376 mm and 800 mm, respectively and Parramatta has an average annual rainfall value (963 mm) very close to the mean value (1048 mm) over the ten locations. It is found that the average annual water savings (AWS) for 'toilet, laundry and irrigation use' in Greater Sydney is strongly correlated with the average annual rainfall (AAR), which can be expressed by the following empirical equations:

$$AWS_2 = 33.56 + 0.0067 \times AAR, r = 0.49 \quad (7)$$

$$AWS_3 = 40.41 + 0.0085 \times AAR, r = 0.50 \quad (8)$$

$$AWS_5 = 44.95 + 0.0162 \times AAR, r = 0.63 \quad (9)$$

where AWS_2 is average annual water savings (kL) for toilet, laundry and irrigation usage for a 2 kL rainwater tank and AAR is average annual rainfall value (mm). AWS_3 and AWS_5 represent 3 kL and 5 kL rainwater tanks, respectively and r represents the Pearson correlation coefficient.

Table 1. Water savings for different rainwater tank sizes considering toilet and laundry usage

Size (kL)	Camp-belltown	Hornsby	Manly	Parramatta	Penrith	Richmond	Sydney	Average
2	27	32	32	30	29	28	32	30
3	30	33	33	33	31	31	33	32
5	33	34	34	34	33	33	34	34

Tables 2, 4 and 6 present reliability values of 2 kL, 3 kL and 5 kL rainwater tanks for all the three cominations of usages considered in this study. It is found that on average a 2 kL rainwater tank can meet the demand for toilet and laundry usage for 86% of the days in a year, which increases to 97% for a 5 kL tank size. It is found that Hornsby has the highest reliability (99% for a 5 kL tank) and Campbelltown has the lowest value. This is well connected with the average annual rainfall values at these two locations (1325 mm and 743 mm for Hornsby and Campbelltown, respectively). It is also noticed that the differences in reliability across different locations reduce with increased tank size. These results suggest that in the peri-urban Sydney districts where Sydney Water supply is unavailable, a minimum tank size of 5 kL is needed, which is expected to meet 95% to 99% of toilet and laundry demands plus the drinking water need (since the drinking water is only a fraction of total water demand in a household). However, if irrigation is considered, 38% to 68% demand can only be met thus suggesting a bigger tank size for a higher reliability.

Table 2. Reliability (%) for different rainwater tank sizes considering toilet and laundry usage

Size (kL)	Camp-belltown	Hornsby	Manly	Parramatta	Penrith	Richmond	Sydney	Average
2	79	92	91	86	82	81	91	86
3	87	96	96	93	89	89	96	92
5	95	99	99	98	96	95	99	97

Table 3. Water savings considering irrigation use only

Size (kL)	Camp-belltown	Hornsby	Manly	Parramatta	Penrith	Richmond	Sydney	Average
2	39	25	31	27	38	25	24	30
3	46	33	41	36	46	33	32	39
5	57	45	56	48	57	44	44	51

Table 4. Reliability (%) considering irrigation use only

Size (kL)	Camp-belltown	Hornsby	Manly	Parramatta	Penrith	Richmond	Sydney	Average
2	36	69	60	56	48	53	67	56
3	36	69	61	58	48	54	67	56
5	38	72	64	61	51	56	70	59

Table 5. Water savings (kL) considering toilet, laundry and irrigation usage

Size (kL)	Camp-belltown	Hornsby	Manly	Parramatta	Penrith	Richmond	Sydney	Average
2	37	40	44	38	39	35	39	39
3	46	48	54	47	48	42	47	47
5	58	61	69	59	61	54	58	60

Table 6. Reliability (%) considering toilet, laundry and irrigation usage

Size (kL)	Camp-belltown	Hornsby	Manly	Parramatta	Penrith	Richmond	Sydney	Average
2	33	64	54	51	43	47	61	51
3	36	67	58	54	46	50	64	54
5	38	69	61	57	49	52	66	57

Table 7. Comparison of water savings across various locations in Sydney considering toilet, laundry and irrigation usage

Location	Water savings for 2 kL tank (kL/year)	Location	Water savings for 3 kL tank (kL/year)	Location	Water savings for 5 kL tank (kL/year)
Manly	45	Manly	55	Manly	71
Penrith	44	Penrith	54	Penrith	69
Cronulla	43	Cronulla	53	Cronulla	68
Parramatta	40	Parramatta	48	Parramatta	61
Hornsby	41	Hornsby	49	Hornsby	61
Sydney	39	Sydney	48	Sydney	61
Bankstown	39	Bankstown	47	Bankstown	59
Kellyville	38	Kellyville	46	Kellyville	59
Campbelltown	42	Campbelltown	52	Campbelltown	65
Richmond	35	Richmond	43	Richmond	54

It was also examined how frequent a tank will remain empty at different locations in Sydney. It was found that a 5 kL rainwater tank for 'toilet, laundry and irrigation' usage in Hornsby is expected to remain empty, more than half-full, and full for 114, 211 and 88 days per year on average, respectively. Figure 5 shows that a tank in Hornsby is expected to remain empty during the months of July to September (when the average monthly rainfall is the lowest in Sydney as shown in Figure 4) and most likely to be full during January to March (when the average monthly rainfall values are the highest in Sydney). It is noted from Figure 5 that a rainwater tank in Hornsby is expected to be at least half full for more than half of the month for all the months except for July, August and September. It was found that Campbelltown performs at the poorest level where a 5 kL rainwater tank remains empty for 227 days per year and full for only 38 days per year on average. These results suggest that rural properties around Campbelltown area of Sydney (where Sydney Water supply is unavailable) needs a rainwater tank larger than 5 kL to ensure an efficient water supply.

The following points may be highlighted from the case study. The reliability of a rainwater tank for toilet and laundry use is similar throughout all the locations considered due to the low water demand for these usages. Here reliability means what percentage of days in a year RWHS can meet the specified demand without the need of mains water. This reliability is very important when RWHS is to be used as a means of sole water supply, e.g. many semi-urban locations in Sydney do not have mains water supply (e.g. part of Hawkesbury City Council). Reliability for irrigation and combined use (toilet, laundry and irrigation) varies remarkably from location to location.

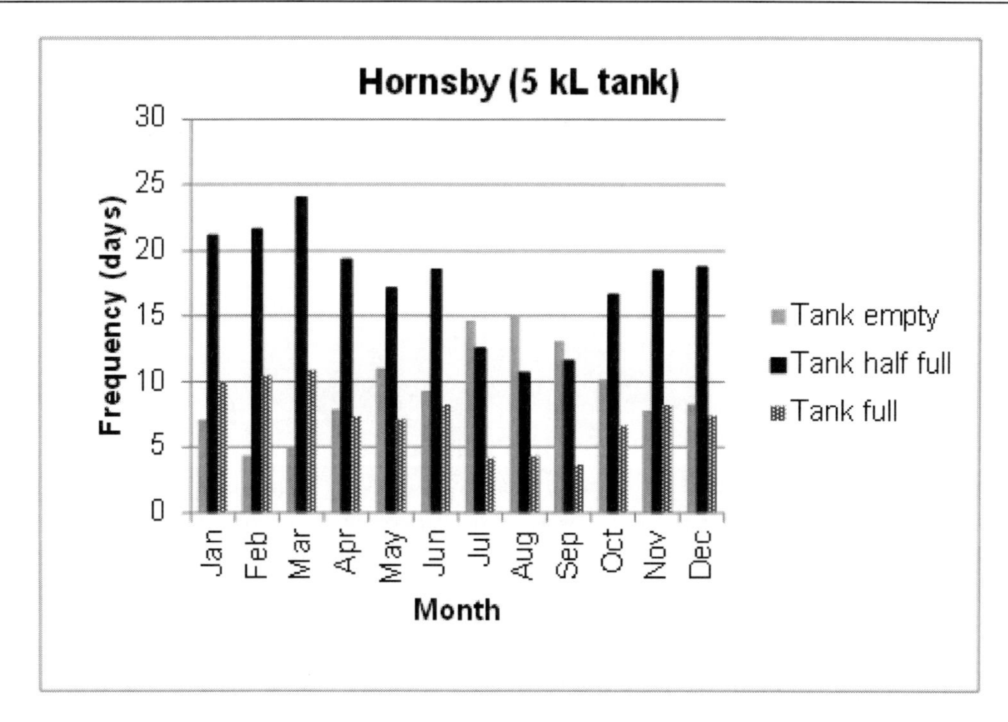

Figure 5. Frequency of rainwater tank being full, half-full and empty at Hornsby (toilet, laundry and irrigation usages).

This is due to the fact that irrigation demand is dependent on daily rainfall, which varies significantly from location to location. Also irrigation demand is higher than toilet and laundry usage.

The most reliable location for all the three usages from all the locations examined is Hornsby. When a 5 kL rainwater tank is used in Hornsby the reliabilities for toilet & laundry, irrigation and combined uses (toilet, laundry and irrigation) are 99%, 72% and 69%, respectively. Among the locations and tank sizes examined, the highest reliability is achieved with a 5 kL tank as expected.

It is found that the increase in reliability from a 2 kL to 3 kL tank is as little as 1% for irrigation and combined use (toilet, laundry and irrigation). A 3 kL or 5 kL tank should be selected for these uses as they have a higher reliability and there is a large increase in water savings compared to a 2 kL tank.

Water savings for toilet and laundry usage are similar across all the locations. When a 5 kL rainwater tank is used, 9 of the 10 locations tested show water savings between 33 kL/year to 35 kL/year.

Water savings for irrigation and combined usage (toilet, laundry and irrigation) are considerably higher than toilet and laundry uses. For irrigation use, the highest water saving location is Penrith with 57 kL/year saving. For combined use the highest water saving location is Manly with 69 kL/year saving.

It is found that a rainwater tank is expected to be empty with the highest frequency during July to September and to be full during the months of January to March. The results show that the average annual water savings from rainwater tanks are strongly correlated with average annual rainfall. Three empirical relationships are developed which can be used to estimate water savings at an arbitrary location in Sydney from the average annual rainfall. These equations can be used in estimating the rainwater tank adequacy for a given location in Sydney. It is also found that 5 kL tank is preferable to the currently recommended 3 kL tank. It is also found that to achieve the best results each Sydney Council should develop their own design curves to identify the best tank size. It is also found that where Sydney Water supply is not available, a minimum of 5 kL tank will be able to meet about 95% of water demand if irrigation is not needed. For irrigation use, a much bigger tank is needed.

CONCLUSION

Rainwater tank has emerged as an important water conservation and recycling element in urban areas of Australia. Due to greater environmental awareness and government rebate, the use of rainwater tanks in Australia has been increasing notably. From the literature it is hard to make any conclusion whether the rainwater tank is economically viable to house owners, which is mainly due to low price of urban water supply. It is thus recommended that rebate for rainwater tank should be increased to make rainwater tanks financially viable to house owners. This will help to achieve a sharp rise in the uptake of rainwater tanks. From a case study in Sydney, it has been found that for a typical detached house, a 5 kL tank is more suited than the currently recommended 3 kL tank. Also, where Sydney Water supply is not available, e.g. in peri-urban regions of Sydney (such as rural areas adjacent to Richmond, Penrith, Hornsby and Cambelltown), a 5 kL tank will be enough to meet the need for toilet, laundry and drinking water demands for about 95% of time.

REFERENCES

Australian Bureau of Statistics (ABS). (2007). Environmental Issues: People's Views and Practices, Available at http://www.abs.gov.au.

ABS (2010). More Australian using rainwater tanks. Media Release, 19 Nov 2010 by ABS. http://www.abs.gov.au.

Campisano A., Modica C. (2012). Optimal sizing of storage tanks for domestic rainwater harvesting in Sicily. *Resources, Conservation and Recycling*, 63, 9-16.

Domenech L, Sauri D. (2010). A Comparative appraisal of the use of rainwater harvesting in single and multi-family buildings of the metropolitan area of Barcalona (Spain): social experience, drinking water savings and economic costs. *Journal of Cleaner Production* 11, 1-11.

Eroksuz E., Rahman A. (2010). Rainwater tanks in multi-unit buildings: A case study for three Australian cities. *Resources, Conservation & Recycling*. 54, 1449-1452.

Ghisi E., da Fonseca T., Rocha V.L. (2009). Rainwater harvesting in petrol stations in Brasilia: Potential for potable water savings and investment feasibility analysis. *Resources, Conservation & Recycling*, 54, 79-85.

Ghisi E., Schondermark P. N. (2013). Investment feasibility analysis of rainwater use in residences. *Water Resources Management*, DOI 10.1007/s11269-013-0303-6.

Imteaz M.A., Rahman A., Ahsan A. (2012). Reliability analysis of rainwater tanks: A comparison between South-East and Central Melbourne. *Resources, Conservation & Recycling*, 66 (2012), 1-7.

Imteaz M.A., Ahsan A., Naser J., Rahman A. (2011a). Reliability analysis of rainwater tanks in Melbourne using daily water balance model. *Resources, Conservation & Recycling*, 56, 80-86.

Imteaz M.A., Shanableh A., Rahman A., Ahsan A. (2011b). Optimisation of rainwater tank design from large roofs: A case study in Melbourne, Australia. *Resources, Conservation & Recycling*. 55, 1022-1029.

Khastagir A, Jayasuriya N. (2009). Optimal sizing of rain water tanks for domestic water conservation. *Journal of Hydrology*, 381, 181-188.

Khastagir A, Jayasuriya N. (2011). Investment evaluation of rainwater tanks. *Water Resources Management*.25, 3769-84.

Kyoungjun K, Chulsang Y. (2009). Hydrological Modeling and Evaluation of Rainwater Harvesting Facilities: Case Study on Several Rainwater Harvesting Facilities in Korea. *Journal of Hydrological Engineering*, 14(6), 545-61.

Mehrabadi M.H., Saghafian B, Fashi F.H. (2013). Assessment of residential rainwater harvesting efficiency for meeting non-potable water demands in three climate conditions. *Resources, Conservation & Recycling*, 73, 86-93.

Muthukumaran S, Baskaran K, Sexton N. (2011). Quantification of potable water savings by residential water conservation and reuse – A case study. *Resources, Conservation and Recycling*. 55, 945-52.

Rahman A., Dbais J., Imteaz M.A. (2010). Sustainability of RWHSs in Multistorey Residential Buildings. *American Journal of Applied Science*, 1(3), 889-898.

Rahman A., Dbais J., Islam S. K., Eroksuz E., Haddad, K. (2012a). Rainwater harvesting in large residential buildings in Australia. In "Urban Development", ISBN 978-953-307-957-8, Editor: Dr. Serafeim Polyzos, Published by InTech, www.intechweb.org, 21pp.

Rahman A., Keane J., Imteaz M. A. (2012b). Rainwater harvesting in Greater Sydney: Water savings, reliability and economic benefits, *Resources, Conservation & Recycling*, 61, 16-21.

Su M, Lin C, Chang L, Kang J and Lin M. (2009). A probabilistic approach to rainwater harvesting systems design and evaluation. *Resources, Conservation & Recycling*, 53, 393-99.

Tam V.W.Y., Tam L., Zeng S.X. (2010). Cost effectiveness and trade off on the use of rainwater tank: an empirical study in Australian residential decision-making. *Resources, Conservation and Recycling*, 54, 178-86.

Van der Sterren M., Rahman A., Dennis G. (2012). Rainwater harvesting systems in Australia. In Water Quality, ISBN 979-953-307-638-5, Editor: Dr Kostas Voudouris, Published by InTech, www.intechweb.org, 26 pp.

Zhang Y., Chen D., Chen L., Ashbolt, S. (2009). Potential for rainwater use in high-rise buildings in Australian cities. *Journal of Environmental Management*. 91, 222-26.

Zhang, X., Hu, M., Chen, G., Xu, Y. (2012). Urban rainwater utilization and its role in mitigating urban waterlogging problems—A case study in Nanjing, China, *Water Resources Management*. 26, 3757-3766.

In: Water Conservation
Editor: Monzur A. Imteaz
ISBN: 978-1-62808-993-6

Chapter 3

WATER SUPPLY THROUGH RAINWATER HARVESTING IN URBAN AREAS

Mohammad Hossein Rashidi Mehrabadi[1*] and Fereshte Haghighi Fashi [2]

[1]Department of Technical and Engineering, Science and Research Branch, Islamic Azad University, Tehran, Iran;
[2]Faculty of Agricultural Engineering and Technology, University of Tehran, Tehran, Iran.

ABSTRACT

In many countries around the world where access to water resources to supply potable water for residents is lacking, rainwater would be a desirable resource to meet potable water demand. Using rooftops rainwater harvesting systems would help to meet water demand in urban areas especially in arid and semi-arid areas. The problem of water supply in urban areas, the high operation and maintenance costs of other resources, and the high costs of developing water resources, especially in arid areas have become a matter of concern. It is feasible to harvest and store surface runoff in residential buildings by rainwater harvesting systems, in order to reduce the volume of surface runoff into the sewer networks, prevent traffic problems during rainfall, and to minimize the surface contaminant transport in the cities. This chapter introduces rainwater harvesting systems and all the factors influencing their performance. This chapter examines and assesses the applicability and challenges of rainwater harvesting for daily non-potable water supply from roofs of residential buildings in three urban areas of Iran. Reliability of the system is investigated in terms of size of roof area, tank size, and water demand. According to the results presented in this chapter, it could be stated that if the volume of tanks and daily demand of residents be calculated based on physical condition of buildings and hydrologic condition of the region, the rainwater storage in tanks will increase and make it possible to supply more water for residents.

* Corresponding Author address: Email: hossein_hakim@yahoo.com & Tel: +98 21 77072309.

Keywords: Rainwater harvesting, residential areas, water demand, hydrological conditions

INTRODUCTION

Nowadays, water supply has become an important issue all over the world. With increasing residential buildings and population in the urban areas, the water demand is increased. The problem of water supply in urban areas, the high operation and maintenance costs of other resources, and the high costs of developing water resources, especially in arid areas have become a matter of concern. Rainwater harvesting is a promising technology for alternative water resources. In many developing countries, rainwater harvesting is important for both economic and social development that can improve living conditions.

In urban areas, most of the ground is covered with buildings and pavements and most of the rainwater which falls in urban areas flows into sewage network, water drains and runs away. Roof rainwater harvesting in urban areas for drinking, non-potable and also irrigation consumptions is a practical solution to reduce increasing crisis in the water supply for citizens. Rainwater harvesting and storage of water for use in future is a viable option to water conservation in urban areas making it possible for residents to meet their demand. In addition to supply the water demands of residents, the collection of rainwater in tanks prevents flood producing and decreases the flood risk in the case of intensive rainfall in urban areas.

This chapter examines and assesses the applicability and challenges of rainwater harvesting for daily non-potable water supply from roofs of residential buildings in three urban areas of Iran. Reliability of the system is investigated in terms of size of roof area, tank size, and water demand.

REVIEW OF LITERATURE ON RAINWATER HARVESTING IN RESIDENTIAL AREAS IN RECENT YEARS

Ghisi et al. (2007) have evaluated the potential of water saving from rainwater harvesting system in Brazil and showed that average potential for potable water savings ranged from 12 - 79% per year for the studied cities. The result of their research showed the average potential for potable water savings was 41% in south-eastern Brazil. They found that rainwater tank capacity should be determined for various locations.

Song et al. (2009) have studied applicability of rainwater harvesting as a viable option for water supply in Banda Aceh city in Indonesia. They discussed the installation methods, public perceptions, and viability of the rainwater harvesting systems and claimed that these systems, accompanied by an increase in public awareness and appropriate training, can help to supply potable water for residents.

Abdulla and Al-Shareef (2009) have evaluated the potential of potable water storage using rainwater harvesting in residential areas of 12 Jordanian governorates. Their research provides suggestions for improvement of quality and quantity of harvested rainwater and shows the importance of rooftop rainwater harvesting systems for domestic water supply. The results of their study showed that the rainwater harvested from the roofs of buildings has mineral compounds in drinking water standards but bacteriological parameter is above the standard. Sturm et al. (2009) have conducted a research on rainwater harvesting in central northern Namibia. They presented and evaluated a appropriate solution for rainwater harvesting based on technical, hydrological, and socio-cultural conditions. They assessed the feasibility of the rainwater harvesting systems considering local socio-economic conditions. According to results of these studies and based on the physical conditions, an appropriate rainwater harvesting design and required storage capacities were presented. Labour costs and Local material were assessed and the cost analysis was done. In overall, results have demonstrated that it is economically feasible to apply techniques of rainwater harvesting in terms of a roof's catchment system.

Ghisi et al. (2009) investigated rainwater harvesting as a viable way to optimize water resources uses. The results of their study showed that the average potential storage of potable water using rainwater is 32.7%. Eroksuz and Rahman (2010) have investigated water saving capability of rainwater tanks in multi-unit buildings fitted in Sydney, Newcastle and Wollongong cities. They stated that for multi-unit buildings a larger tank size was the most appropriate to maximize water saving. They also found that these kinds of tanks supplied significant water savings even in dry years. They developed a prediction equation to estimate average annual water savings in above mentioned three Australian cities.

Khastagir and Jayasuriya (2010) investigated the reliability of rainwater tanks and presented design charts of optimum tank sizes regarding data on daily rainfall, rainwater demand, and roof area required to supply 90% reliability demand. Jones and Hunt (2010) evaluated the use of rainwater harvesting systems in North Carolina. They have developed a computer model to simulate system performance. Results of their study showed that the improved performance of large systems providing an indication of reduced returns for increased cistern capacity.

Domenech and Sauri (2011) have evaluated the use of rainwater harvesting systems in residential buildings based on social experience, saving drinkable water and economic value in Barcelona in Italy. They showed that based on low precipitation and in order to have equity between the system and demand, it's possible to use the low volume tanks. The main problem of this method is that cost recovery for rainwater harvesting systems is high for residential buildings and on the other hand, if families use this method for systems, it will bring some advantages for the society.

Palla et al. (2011) have assessed optimum performance of a rainwater harvesting system and developed a model for determining the efficiency of a water-storage system, over flow ratio and detention time. Performance was assessed under various circumstances in terms of environmental conditions as means of three precipitation regimes and three levels of water demand. Imteaz et al. (2011a) have studied the optimization of rainwater harvesting tanks from large area roofs in Melbourne in Australia. Their study demonstrated the need for detailed cost - benefit analysis and optimization for large rainwater tanks to maximize the benefits. Imteaz et al. (2011b) attempted to develop a comprehensive decision support tool for the performance analysis and design of rainwater tanks in Melbourne. They presented several reliability charts for domestic rainwater tanks in relation to tank volume, roof area, and numbers of people in a house and percentage of total water demand to be supplied by harvested rainwater for three climatic conditions. Their results showed that situations with a household of two occupants, it was possible to achieve about 100% reliability with a roof size of 150–300 m^2 and a tank size of 5000–10,000 L.

Imteaz et al. (2012) have evaluated the potential of rainwater in southwest of Nigeria, on the basis of physical and climatic conditions and daily water supply. They stated that the increasing volume of rainwater storage tanks could help to collect and meet the water demands of residents. Based on studies conducted in this research, to achieve 100% water demands of residents, 7000 L tanks' volume is required and to supply water for many residents, 10,000 L tanks' volume is needed. Rahman et al. (2012) have investigated the water storage potential of rainwater tanks at the various locations in Sydney, Australia. They developed a daily water balance simulation model on and water storage, reliability and financial aspect were estimated for tank sizes of 2 kL, 3 kL and 5 kL. They found that the average annual water storage in rainwater tanks was correlated with average annual rainfall. Moreover, it was found that the cost benefit for the rainwater tanks were lower than 1.00. According to the results, it was found that the 5 kL tank is better than 2 kL and 3 kL tanks. It was noted that if rainwater tanks be connected to toilet, laundry and outdoor irrigation, the financial outcome for the home owners will be the best. The results of their study showed that government authority has an important role to enhance rainwater harvesting acceptance by using maintaining or increasing the rebate for rainwater tanks. Rashidi Mehrabadi et al. (2013) assessed the applicability and performance of rainwater harvesting systems to supply daily non-potable water. They simulated storage of rainfall on the roofs of residential buildings in three climatic conditions, Mediterranean, humid, and arid climate. They found that in humid climate, it was possible to supply at least 75% of non-potable water demand by rainwater storage from larger roof areas (for a maximum duration of 70% of the times). For Mediterranean climate, they concluded it was possible to supply at least 75% of non-potable water demand in buildings with larger roof areas (for a maximum duration of 40% of the times). Finally, it was concluded that in arid climate, similar duration is only 23% of the times.

Rainwater Harvesting Systems in Buildings, Advantages, and Limitations

In many regions of the world, there is water availability problem and solving this problem needs to large investments. Rainwater is an available and relatively clean resource that can be used as a source of drinking water. The rainwater harvested from roofs is usually free and cleaner than water collected from other resources. Rooftop rainwater harvesting systems could provide drinking water with a high quality and reduce stress on drainage networks and canals. In addition, these systems can prevent soil erosion and groundwater recharge. Rainwater harvesting may reduce urban water consumption and consequently reduce the cost of water purchasing. In some regions, the local residents spend a more percent of their incomes on the purchase of water. For this reason, a economical, novel, and sustainable option of water supply is a most important concern. Rainwater harvesting systems can be implemented in new buildings as well as old buildings. Rainwater offers considerable advantages that can be used for various purposes such as drinking and irrigation. The implementation of rainwater harvesting technologies would be feasible and easy for the following reasons:

- These systems are applicable at small-scale with a minimum of technical knowledge,
- These systems have advantages especially in operation performance,
- They are economically feasible over the other sources,
- They improve the drainage conditions of the urban areas,
- They are adaptable to a variety of climate and physical conditions.

Rainwater harvesting can be used in a variety of purposes regardless of water quality. Furthermore, these systems have important environmental, economical, and cultural effects.

However, it should be considered that water harvesting from roofs has some disadvantages. Poorly managed system may leads to the other problems. Some major disadvantages of rainwater harvesting from roofs are as follow:

- Air pollution, dust and debris, bird droppings, insects, dirt, and organic matter may influence rainwater quality,
- The amount of rainfall and the roof area and tank size are limiting factors,
- Droughts can cause water supply problems,
- Maintenance and operation problems are important considerations.

It is needed to systems for treating water during storage, which help to manage water quality and prevent water contamination. The most important parameters influencing design of roof rainwater harvesting systems are including daily rainfall statistic of the region, rainwater harvesting area (roof), runoff coefficient, tank size, costs, tank installing and maintenance, and the daily water needs of residents. Proper design of rainwater harvesting systems in residential areas can lead to provide the potable and non-potable water for residents and flood control in urban areas and reduced emissions of contamination in cities. These systems are implemented in areas having a high quality daily precipitation data leading to have more cost effective systems and more storage of rainwater. The rainwater storage tank needs to the more investment than the other components of rooftop rainwater harvesting system. Therefore, accurate design of tank is necessary to supply optimum storage volume while being the cost effective. It is not possible to estimate the implementation cost of rainwater harvesting systems accurately because it depends on a variety of factors including the rooftop surface, tanks, pipes, and other materials and components.

Finally, to implement the rainwater harvesting systems in residential buildings, it's feasible to get help from government agencies, people, and NGOs. For implementation of rainwater harvesting systems, local people can be trained to reduce costs and increase more participation.

In general, it is important developing rainwater harvesting systems in areas where the potable water supply is limited. Rainwater harvesting is a viable option for water supply in terms of cost and maintenance as well as water quality.

There are some technical shortcomings facing rainwater harvesting implementation including the lack of design guidelines for tanks, storage tank size with respect to the climate and costs, and the lack of water quality control and system maintenance. Therefore, it is important to assess technical considerations and performance of rainwater harvesting systems for non-potable water supply from roofs of residential buildings in different climatic conditions. In addition, reliability of the rainwater harvesting systems should be investigated in terms of roof area, tank size, and water demand. Reliability is the percentage of days in a year when rainwater tank is able to supply the intended partial demand for a specific condition (Imteaz et al. 2011b).

Rainwater harvesting systems collect and store the rainwater runoff from roof of the buildings and from the ground. Rainwater harvesting system consists of three components including the rainwater harvesting area (roofs of buildings), water transfer system and tank. The rainwater harvesting process includes rainwater catching, directing rainwater to tank, and filtering it for different consumptions immediately or for future use during days of water shortage.

Figure 1 shows the rainwater harvesting system on the roof of the building schematically.

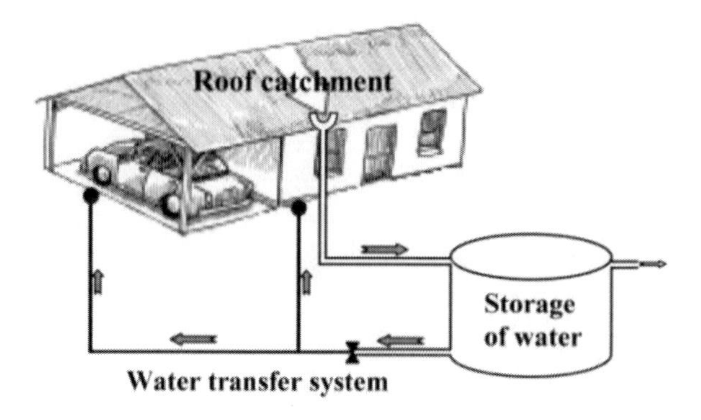

Figure 1. Schematic view of rainwater harvesting system in residential buildings.

Rainwater Harvesting Area (Roof of Building)

The catchment in rainwater harvesting system is the surface catching the rainwater which causes the collection of rainwater into the water transfer system. Rooftops of residential buildings are important as rainwater harvesting surface because a large volume of rainwater is collected easily from these surfaces. Roof size of a building would determine catchment and runoff from rainwater. In the case of arid regions, if residential building roof area be not large enough for rainwater collection in tanks, so can not meet the water demands of residents. In this condition, the other resources should be considered to provide water supply. Therefore, in such regions the roof area should be large enough to harvest and save considerable volume of rainwater in tanks. Runoff coefficient of roof of buildings is an effective parameter for rainwater harvesting. A more runoff coefficient leads to more rainwater harvested. An impermeable roof leads to more runoff and also roofs having small runoff coefficients would cause less runoff. During rainwater falls in rooftops, the rainwater could be contaminated not only due to the materials used in the roof, but also through materials that are accumulated on the roof. Periodical cleaning of the roof surface in residential buildings could be helpful to reduce roof buildings contamination. However, keeping clean the rainwater harvesting surface (roof of buildings) is not easy possible and is time and finance consuming.

Rainwater Transfer System (Gutters)

Water harvested from rooftops should be transferred to the storage tank through a system of gutters and drains. There are several types of transfer system, but gutters are the best and less cost option used in residential buildings. Materials used in Gutters and pipes attached to the gutters are used commonly in the form of galvanized metal or plastic tubes (made from P.V.C.) which are available in the market. The performance of roof gutters should be such that the water on roof surface totally be transferred without overflow to water storage tank. The roof gutters should be resistant against various climatic conditions such as storm rain, snow, and sunlight. Connections between water transfer networks should be resistant in order to water transfer without any problem.

Storage Tank

The main component of rainwater harvesting systems is construction or purchase of tanks. Volume of tanks has a direct effect on the volume of stored rainwater and water overflow from tanks. Proper and optimum design of tanks' volume is the most important issue in design of rainwater harvesting systems. Optimum design of tanks and optimized volumes of tanks can be achieved by modeling and simulation by optimization programs based on the region and building conditions. If the size of tanks not be designed considering buildings and climate conditions, systems would have not efficiency and could not make proper storage of rainwater in the region. The water storage tanks are more influential component in rainwater harvesting systems in residential areas. Hence, to make an optimum storage capacity and maintain rainwater in residential buildings requires accurate planning and this work should be done with the least cost and time. Materials used for tank construction are including concrete, plastic, metal sheets, wood and etc. Determination of size of the storage tank is the most important part of the design phase in the project. There are several methods that can be used to determine the volume of tanks. Six parameters are important in the calculation of the tanks size for optimum design of the rainwater harvesting tanks in residential buildings including:

1) Daily rainfall statistics of the region.
2) Rainwater harvesting area (roof area).
3) Water demand of residents.
4) Construction and installation costs of tanks.
5) Maintenance and repair costs of tanks and their lifetime.
6) Availability of materials to construct the tanks.

Among all types of tanks, usually the most appropriate and the most common are plastic polyethylene ones. Compared to the other tanks, this type of tanks is of less weight. This leads to great reduction of transportation costs, equipment needed for installation, labor, etc. In addition, the speed of installation of all equipments will be much more. The polyethylene tanks can be installed both buried and also open. These tanks can be installed in horizontal cylinder form or in smaller volumes as a vertical cylinder.

Study Area and Data Used

To investigate the water harvesting systems in residential buildings, residential buildings from Rasht and Gorgan in north of Iran were used (Figure 2) which have temperate and humid climates, respectively. Daily rainfall data for 53 years from 1956 to 2008 were collected for each city. Figures 7 and 8 show the mean daily rainfall for the 53 years from January to December. In Table 1, average rainfall statistics for this period are shown in the two cities. The purpose of implementation of rainwater harvesting systems in these buildings located in cities was supply of non-potable water demand of residents. The non-potable water demand was 80 liters in a day per resident by average and also expectancy level of water supply from rainwater harvesting systems in these tanks is at least 75%. Capacity of rainwater tanks were from 1000 to 20000 L by 1000 L increment. Roof area of buildings in cities was considered 100, 200 and 300 m^2. Runoff coefficient of buildings rooftops were considered 0.85 based on the material and slope of the roof. Residents of the rooftops area 100, 200 and 300 m^2, were selected 4, 8 and 12 people, respectively.

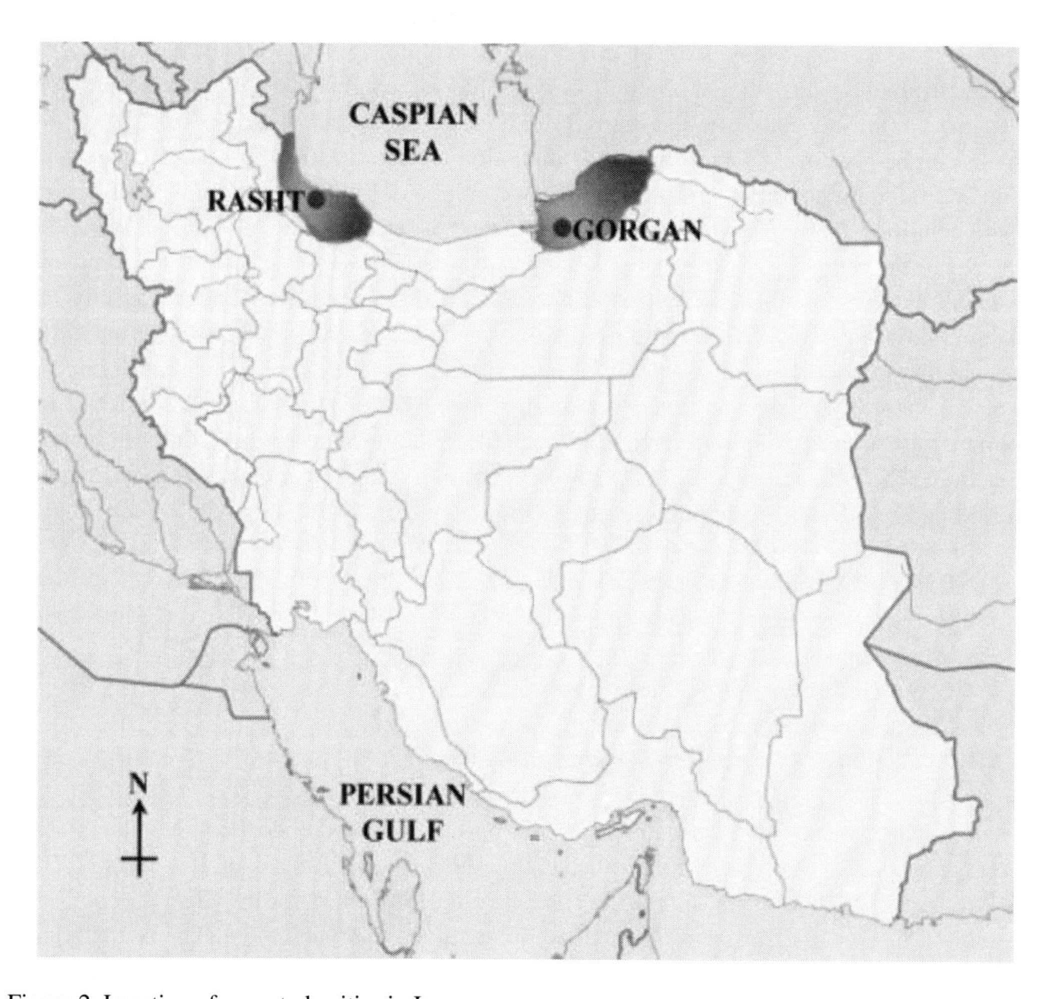

Figure 2. Location of case study cities in Iran.

Table.1. Average annual rainfall in cities

City	Period	Average annual rainfall (mm)
Rasht	1956-2008	1355
Gorgan	1956-2008	607

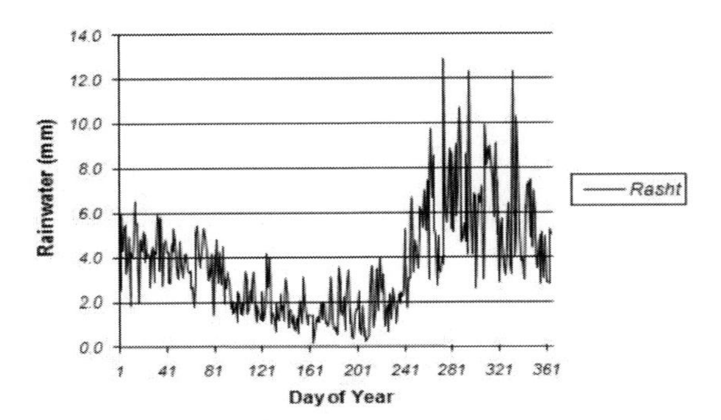

Figure 3. Average daily rainfall depths in Rasht city (Starting from January 1)

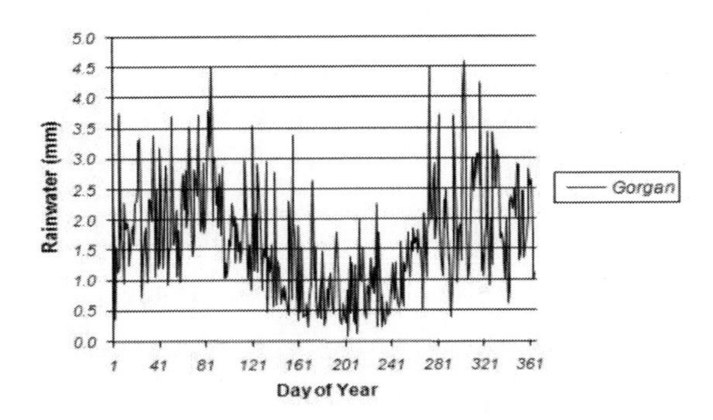

Figure 4. Average daily rainfall depths in Gorgan city (Starting from January 1).

METHODS

Evaluation Method for Rainwater Harvesting Systems

Volume of water stored from the rooftops water harvesting depends on hydraulic parameters of rainwater harvesting system. Parameters affecting system design are including daily rainfall statistics, runoff coefficient of rooftops, roof surface area, number of occupants, and daily water demand of residents. The main component of the rainwater harvesting system in residential buildings is water storage tank designing. The Knowledge of the tanks hydraulic parameters and hydrologic condition of the urban area help to proper and accurate design of tanks in order to optimum design of the volume of these tanks and reduced cost of tanks construction. If the rooftops water harvesting systems be modeled and simulated based on physical and hydrological conditions of the region, and the design and implementation be based on obtained charts and graphs, this cause to increase the efficiency of the systems. According to the previously mentioned notes, for implementation of rainwater harvesting systems in residential buildings, it can be stated that:

1- Rainfall statistics of the studied region affect on system performance. If is the study area have not appropriate rainfall statistics, this leads to the unsuitable performance of the harvesting system and system consequently may not be economically justified. Rainwater harvesting from the roofs of the buildings is common in the rainy areas and not common in areas with low rainfall or arid areas.

2- The area, slope, and type of residential buildings roof affect on the amount of roof rainwater harvesting. A more roof area leads to the much more volume of rainwater harvesting and consequently the more water storage during rainfall. In addition, the materials and slope of the roof affect on runoff coefficient. Runoff coefficient being closer to one, the rainwater harvesting from the roof will increase.

3- Tank volume used in rainwater harvesting system would affect on water storage. If the tank volume be smaller than the roof area and rainfall amount, the system may not perform well. Moreover, if the tank volume be considered larger than the roof area and rainfall amount, this increases the costs of implementation, maintenance, and repair which is economically not justified.

In order to analysis of water harvesting systems to supply water demands of residents, at the first, daily rainfall data should be collected in subsequent years. After collection of rainfall data, the physical characteristics of the buildings such as area, runoff coefficient of the roof, and the water needs of the residents sould be determined. After collecting the initial data, the equations and the system conditions in building should be determined. After determining the system requirements, the system shoul be modeled. After the above mentioned steps, analysis of rainwater harvesting system can be done in order to supply water demands of the residents in the building based on the physical condition of the buliding and hydrological condition of the region. The above mentioned steps can be displayed as Figure 5.

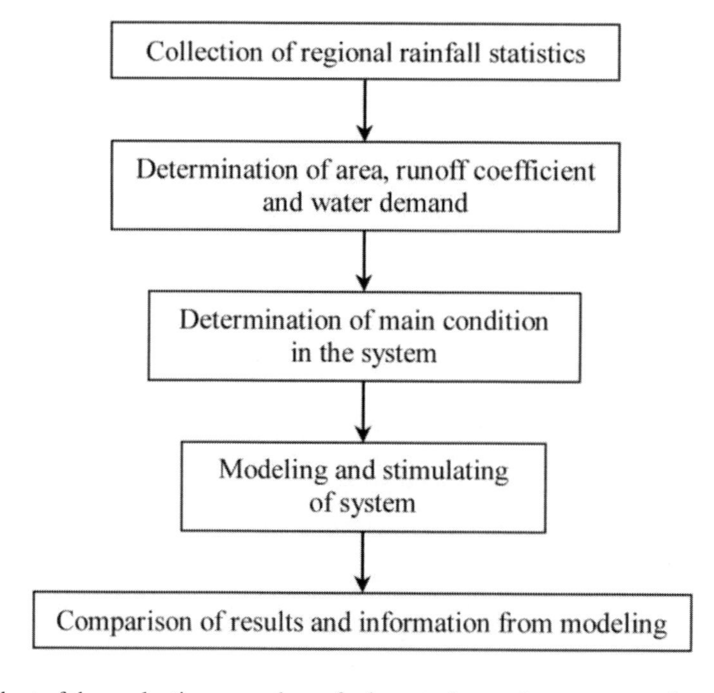

Figure 5. Flowchart of the evaluation procedure of rainwater harvesting system performance.

Model simulation

The following relations can be used to calculate total factors involved in rainwater harvesting systems in residential buildings:

The total volume of rainwater harvestable from roof surface I_t (L) is obtained as follows:

$$I_t = R_t \times A \times \varphi \tag{1}$$

In this equation, R_t is daily rainfall (mm), A is the roof area (m^2), and φ is runoff coefficient of the roof (dimensionless). The volume of the water stored in the tank is calculated with the following equation:

$$V_t = I_t + V_{t-1} - O_t - SP_t \tag{2}$$

where V_t is the water storage volume in time t (L), V_{t-1} is volume of the water stored in tank in the past days (L), SP_t is volume of the tank spillway in time t (L), and O_t is the harvested water from the tank to meet water demands of residents (L) (Figure 6).

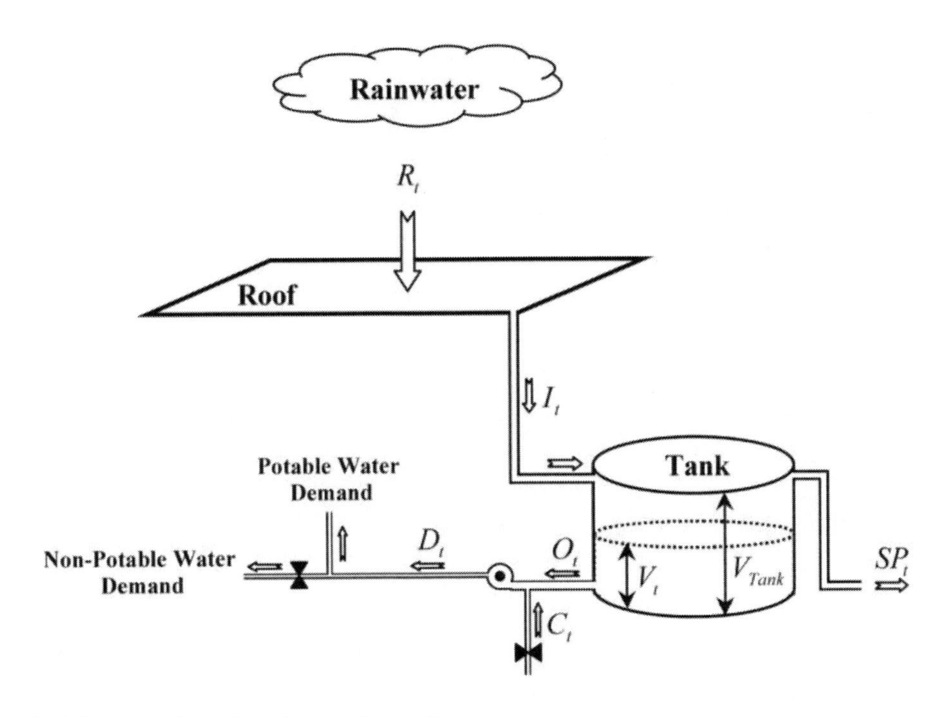

Figure 6. Schematic view of a rainwater harvesting system.

In some cases, the amount of water stored in the tank is less than the daily needs of residents (D_t). To supply this water shortage, urban water (C_t) is used and the volume of water harvested from the tank (O_t)is used to supply water demands of the building residents. The total volume of the water needed to be harvested is calculated with the equation:

$$D_t = C_t + O_t \tag{3}$$

Based on the following equation, C_t can be considered as follows:

$$\begin{aligned} &If: O_t \geq D_t \rightarrow C_t = 0 \\ &If: O_t < D_t \rightarrow C_t = D_t - O_t \end{aligned} \tag{4}$$

The total days of year to supply water for residents is calculated with the equation:

$$\mathrm{Re} = \frac{D_w}{D_d} \times 100 \tag{5}$$

where Re is the reliability of the tank to supply water demand, D_w is the number of days the tank is able to meet the demand for residents, and D_d is the total days of a year. Flow chart used to tank simulation is shown in Figure 7.

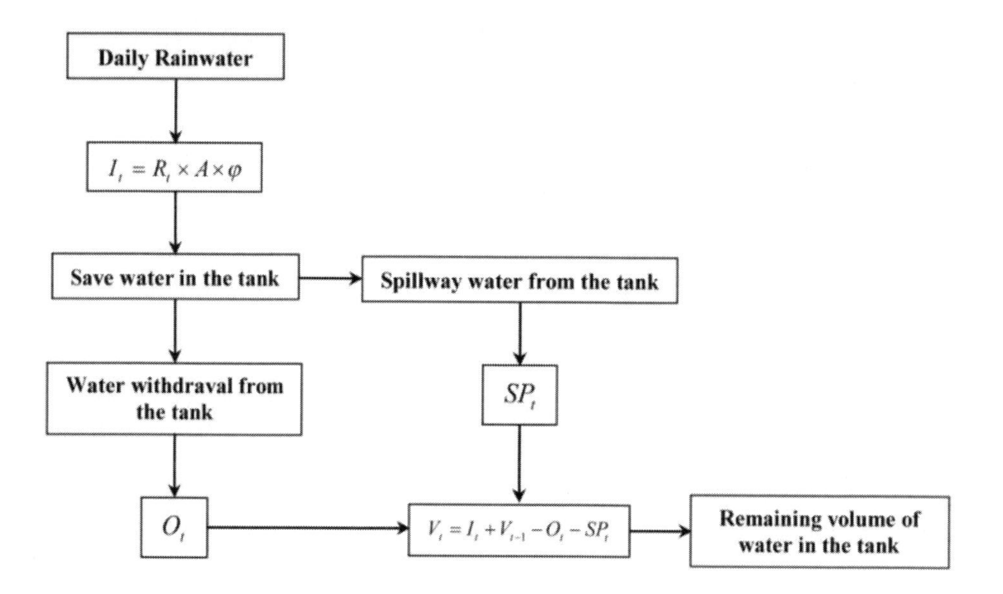

Figure 7. Flow chart used to simulation a rainwater harvesting system.

Results

Rainwater harvesting system from the roof of the building was simulated in the studied cities using Matlab software based on the equations described in before section. For system simulation, the volume of harvestable rainwater from roofs, volume of water stored in tanks, and the percent of water supply (non-potable water) were calculated as daily for the entire period of rainfall statistics in two cities. Figures 8 and 9 show Rasht and Gorgan cities, respectively indicating the percentage of days of year to supply non-potable water to meet the daily needs of urban residents for different tank volumes for the roofs area of 100, 200 and 300 m^2.

According to the figure 8, it can be stated that in Rasht city, the maximum number of days of year which minimum 75% of daily non-potable water demand could be supplied is equal to 82% of total days of year for 100 m^2 of roof area, 320 L non-potable water demand per day, and tank volume of 20000 L. In addition, the minimum number of days of year is 22% for 300 m^2 rooftop area, 960 L non-potable water demand, and tank volume of 1000 L. By increasing the volume of tank, total days of supply non-potable water demand has increased, but with increasing daily non-potable water demand (increasing residents which is according to the rooftop area), the days of water supply have decreased. According to figure 9, it can be stated that in Gorgan city, the maximum total days of water supply for 320 and 640 L non-potable daily water demands and for large tank volumes is equal to 40% of total days of year and minimum total days of water supply for 960 L non-potable daily water demand and also for tank volume of 1000 L is equal to 14% of total days of year. For rooftop area of 100 m^2 and tank volume of more than 8000 L capacity, total days of water supply is constant and equal to 40% of total days of a year. For roof area of 200 and 300 m^2, total days of water supply for tanks volumes of beyond 15000 L are approximately 40% of total days of year.

Figures 10 to 12 show the total days of daily non-potable water supply in Rasht city, for 300 to 1000 L non-potable demands variations, roof areas of 100, 200 and 300 m^2, and the tank volumes of 1000, 5000, 10000, 15000, and 20000 L. Based on the figures, it can be stated that for Rasht city, the maximum efficiency of the system and total days of non-potable water supply is 97% of total days of year for tank's volume of 20000 L, rooftop area of 300 m^2, and 300 L daily non-potable water demand. Also, the minimum efficiency of the system and total days of non-potable water supply is 13% of total days of the year for tank volume of 1000 L, roof area of 100 m^2, and daily non-potable water demand of 1000 L. Increasing the roof area and tanks volume, the total days of water supply would increase and for 300 L non-potable water demand is 39 to 97% of total days of year and for 1000 L non-potable water demand is 13 to 67% of total days of year. In average, the total days of water supply for non-potable water demands of 300 to 1000 L for tank volume of 20000 and 1000 L and roof area of 100 m^2 is equal to 48 and 21% of the total days of year. For 200 m^2 roof area, water supply is equal to 74 and 28% of the total days of year and also for 300 m^2 roof is equal to 83 and 31% of total days of year.

Figures 13 to 15 show the total number of non-potable water supply for Gorgan residents with the same condition in Rasht city. Range of the total days of non-potable water supply for 100, 200 and 300 m^2 roof areas are equal to 6 to 46%, 10 to 84% and 14 to 96% of the total days of a year, respectively.

The average number of supply days for tanks' volume of 1000, 5000, 10000, 15000 and 20000 L for 300 L daily non-potable water supply and for roof area of 100, 200 and 300 m^2 is equal to 39%, 69% and 78% of total days of a year, respectively. This amount for 1000 L non-potable water demand, for roof areas of 100, 200 and 300 m^2 is equal to 9%, 20% and 31%, respectively.

Comparison of reliability for different tank volumes and water demands between the Rasht and Gorgan cities are shown in Table 2. According to Table 2, Rasht stores the highest rain-water volume per year for various roof areas.

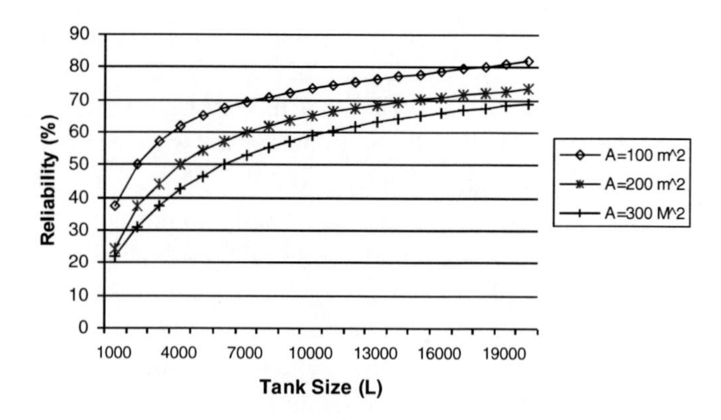

Figure 8. Reliability curves for non-potable water supply for different roof area in Rasht.

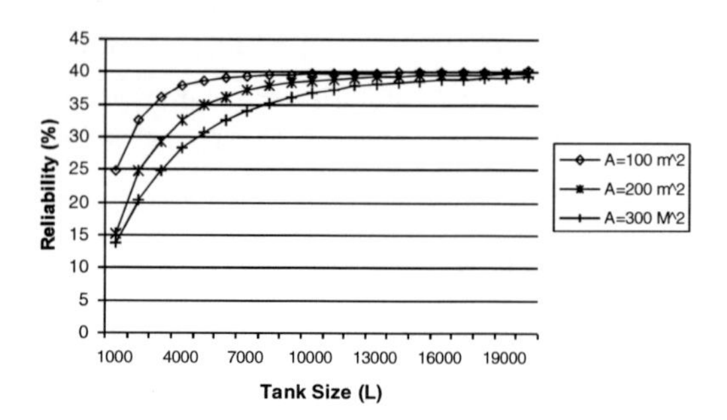

Figure 9. Reliability curves for non-potable water supply for different roof area in Gorgan.

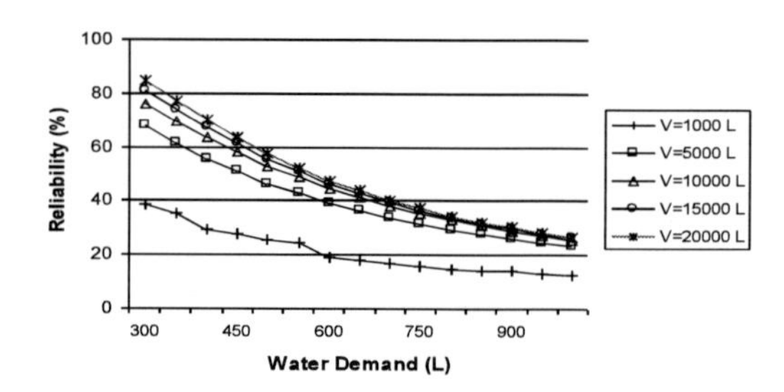

Figure 10. Reliability curves for non-potable water supply for different demands and roof area of 100 m^2 in Rasht.

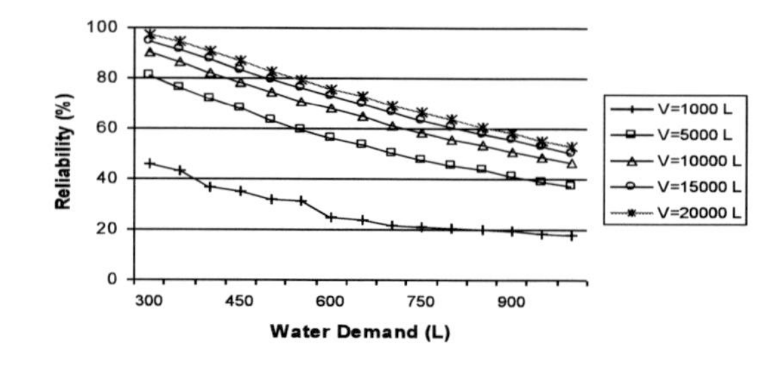

Figure 11. Reliability curves for non-potable water supply for different demands and roof area of 200 m^2 in Rasht.

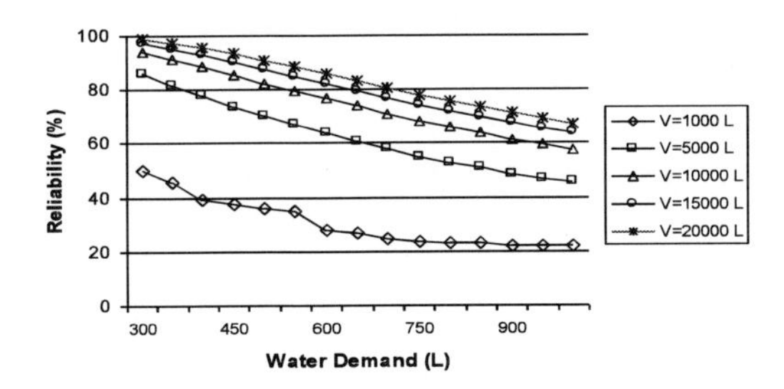

Figure 12. Reliability curves for non-potable water supply for different demands and roof area of 300 m^2 in Rasht.

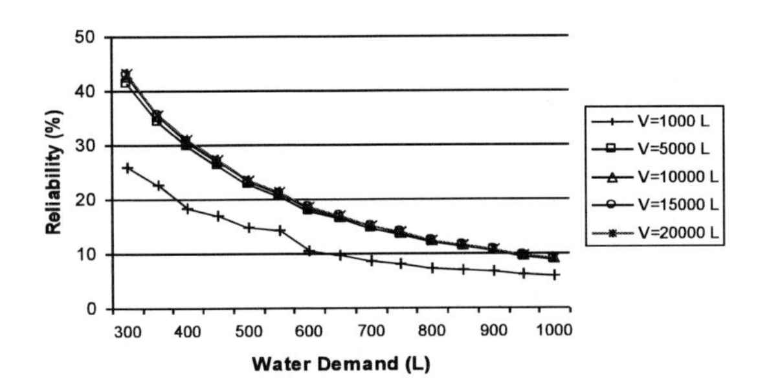

Figure 13. Reliability curves for non-potable water supply for different demands and roof area of 100 m^2 in Gorgan.

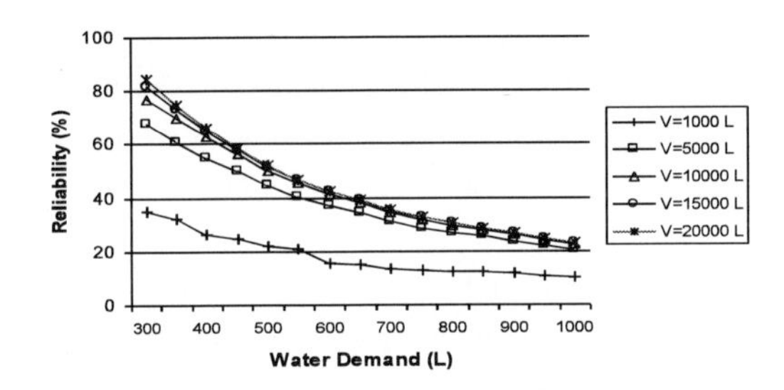

Figure 14. Reliability curves for non-potable water supply for different demands and roof area of 200 m^2 in Gorgan.

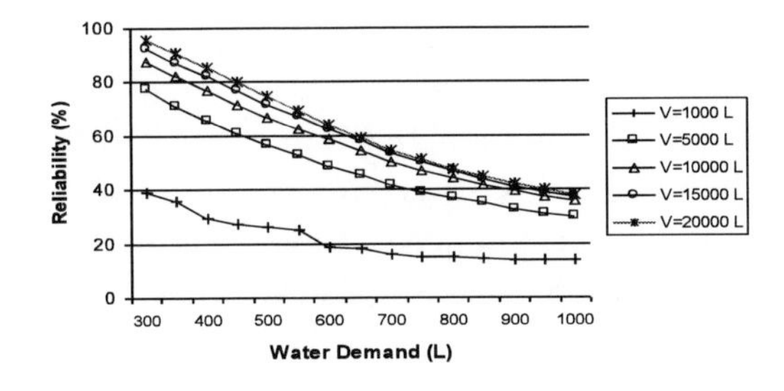

Figure 15. Reliability curves for non-potable water supply for different demands and roof area of 300 m^2 in Gorgan.

Table 2. Comparison of reliability for various demands, roof areas and tank volumes

City	Roof area (m^2)	Demands (L)	Range of reliability (%) for different tank volumes (L)				
			1000 L	5000 L	10000 L	15000 L	20000 L
Rasht	100	300-1000	13 - 39	23 - 68	26 - 76	26 - 81	27 - 85
	200	300-1000	18 - 46	38 - 81	46 - 91	51 - 95	53 - 97
	300	300-1000	22 - 50	38 - 81	46 - 91	51 - 95	53 - 97
Gorgan	100	300-1000	6 - 26	9 - 41	9 - 43	10 - 44	10 - 46
	200	300-1000	10 - 35	21 - 68	23 - 77	23 - 82	23 - 84
	300	300-1000	14 - 39	30 - 77	36 -88	37 -92	38 - 96

Conclusion

In comparison with the other methods, rainwater harvesting from rooftop of residential buildings is a simple and low cost technique to supply water demand of the residents. Harvested rainwater could be supplement for other water resources, especially for regions with a low volume of potable water or ground water storages. It's also a proper alternative under drought condition or falling ground water level. In addition, if the rainwater be harvested from large impervious surfaces of the cities, surface runoff and consequently, flood potential would decrease.

We investigated the performance of rainwater harvesting system for saving of rainwater and supplying water demands of the residents in accordance with the regional and climate condition in Rasht and Gorgan cities having a humid and moderate climate, respectively. It can be concluded that with increasing the daily water demand of residents, the total days of water supply by storage in tanks would decrease and also, with increase of roof area, the total days of water supply for residents would increase. This increase is tangible in large roof surfaces than small roof surfaces. In cities with high and enough rainfall, it's possible to supply significant amount of daily water demand for residents per year, based on regional and climate condition. For cities having low amount of rainfall, it's not possible to save and supply water for all days of a year for residents. In these regions, in order to supply parts of water demand of the residents, the roof area and tanks volume should be considered large enough to have maximum capacity in order to save rainwater in rainy days. For cities having moderate rainfall, in order to supply water to meet the demands of residents, it's necessary to determine daily water demand of residents based on regional and climatic condition of the city and also physical condition of building. In such climate, it's not possible to supply enough water for residents in the case of small roof areas and low volume tanks. In order to determine tank volume, the amount of rainfall and roof area should be considered to achieve optimum benefit out of rainwater tanks. It is suggested that the total days of water supply from rainwater would increase, if the volume of tank and the total amount of daily water demand of residents be determined based on roof area, rainfall statistics, rainfall distribution, and number of days required to water storage, as well.

REFERENCES

Abdulla, F. A., Al-Shareef, A. W. (2009). Roof rainwater harvesting systems for household water supply in Jordan. *Desalination Journal*, 243, 195–207.

Domenech, L., Sauri, D. (2011). A comparative appraisal of the use of rainwater harvesting in single and multifamily buildings of the Metropolitan Area of Barcelona (Spain): social experience, drinking water savings and economic costs. *Journal of Cleaner Production*, 19, 598-608.

Eroksuz, E., Rahman, A. (2010). Rainwater tanks in multi-unit buildings: a case study for three Australian cities. *Journal of Resources Conservation and Recycling*, 54,1449–52.

Ghisi, E., Bressan, D.L., Martini, M. (2007). Rainwater tank capacity and potential for potable water savings by using rainwater in the residential sector of southeastern Brazil. *Journal of Building and Environment*, 42, 1654–66.

Ghisi, E., Tavares D. D. F., Rocha, V. L. (2009). Rainwater harvesting in petrol stations in Brasilia: potential for potable water savings and investment feasibility analysis. *Journal of Resources, Conservation and Recycling*, 54, 79–85.

Imteaz, M. A., Shanableh, A., Rahman, A., Ahsan, A. (2011a). Optimization of rainwater tank design from large roofs: A case study in Melbourne, Australia. *Journal of Resources, Conservation and Recycling*.

Imteaz, M. A., Ahsan, A., Naser, J., Rahman A. (2011b). Reliability analysis of rainwater tanks in Melbourne using daily water balance model. *Journal of Resources, Conservation and Recycling*, 56, 80–86.

Imteaz, M. A., Adeboye, O. B., Rayburg, S., Shanableh, A. (2012). Rainwater harvesting potential for southwest Nigeria using daily water balance model. *Journal of Resources, Conservation and Recycling*, 62, 51-55.

Jones, M. P., Hunt, W. F. (2010). Performance of rainwater harvesting systems in the southeastern United States. *Journal of Resources, Conservation and Recycling*, 54, 623-629.

Khastagir, A., Jayasuriya, N. (2010). Optimal sizing of rain water tanks for domestic water conservation. *Journal of Hydrology*, 381, 181–8.

Palla, A., Gnecco, I., Lanza, L. G. (2011). Non-dimensional design parameters and performance assessment of rainwater harvesting systems. *Journal of Hydrology*, 401, 65–76.

Rahman, A., Keanea, J., Imteaz, M. A. (2012). Rainwater harvesting in Greater Sydney: Water savings, reliability and economic benefits. *Journal of Resources, Conservation and Recycling*, 61, 16– 21.

Rashidi Mehrabadi, M. H., Saghafian, B., Haghighi Fashi, F. (2013), Assessment of residential rainwater harvesting efficiency for meeting non-potable water demands in three climate conditions. *Journal of Resources, Conservation and Recycling*, 73, 86-93.

Song, J., Han, M., Kim, T., Song, J. (2009). Rainwater harvesting as a sustainable water supply option in Banda Aceh. *Journal of Desalination*, 248, 233–240.

Sturm, M., Zimmermann, M., Schütz, K. (2009). Urban W, Hartung H. Rainwater harvesting as an alternative water resource in rural sites in central northern Namibia. *Journal of Physics and Chemistry of the Earth*, 34, 776–85.

In: Water Conservation
Editor: Monzur A. Imteaz
ISBN: 978-1-62808-993-6

Chapter 4

DECISION DILEMMA IN ADAPTING STORMWATER SYSTEMS TO CLIMATE CHANGE: A TALE OF THREE CITIES

Oz Sahin, Raymond Siems, Rudi van Staden and Graham Jenkins*
Griffith School of Engineering
Griffith University, Queensland, Australia

ABSTRACT

Infrastructure assets require substantial capital investment to ensure long operational lifespans. This poses significant risk to local governments when considering that infrastructure is sensitive not only to the climate at the time of construction, but also to climate variations over their lifetime. The impacts of the 2011 Queensland flooding demonstrated the risks that infrastructure can face, and highlighted the economic and social damages associated with these types of events. Specific to this research, stormwater infrastructures are considered one of the most costly and vulnerable to a changing climate. It is therefore critical that stormwater systems can be adapted to withstand current as well as future impacts of natural hazards brought about by climate change. In light of this, this study focuses on identifying and evaluating a set of adaption alternatives to prepare stormwater systems for future climatic changes. To evaluate these alternatives, the authors adopt a Multiple-criteria Decision approach involving multiple stakeholders across three coastal cities in South East Queensland, Australia. A decision hierarchy was formed by consulting three stakeholder groups, which included professionals from engineering, finance and planning sectors of local government. Then, a goal, five criteria and five alternatives were determined. Data collected from the three stakeholder groups were analysed, and the preliminary results indicate that each stakeholder group identifies different priorities for each of the alternatives. The highest priority for the engineering and planning departments was to *Modify Planning and Land Use Control Standards*. However, the finance department identified that to *Change the Stormwater Infrastructure Design Standards* was the highest priority. Furthermore, it was observed that the priority values for the adaptation alternatives varied significantly

* Email: graham.jenkins@griffith.edu.au.

depending on whether these were based on the combined judgements of all participants, or the judgements of the individual stakeholder groups and regional councils. This suggests that future decision makers need to structure a hierarchy model that restricts stakeholder groups to only making priority judgments with regard to themselves.

INTRODUCTION

In recent years, as experienced in Queensland during 2010-11 and 2013, severe storms have caused flooding, resulting in damage to homes and infrastructure including stormwater pipes and roads. In the future, as a result of a changing climate, greater extremes of rainfall, drought and heat waves are expected in South East Queensland (SEQ). Precipitation is likely to be more variable and less predictable with more frequent high intensity rain showers. The risk of negative impacts associated with these events, therefore, is also expected to increase. In recognition of these threats, federal, state and local government bodies across Australia have been developing short, medium and long term mitigation strategies (Siems *et al.*, 2013; Sahin *et al.*, 2013a). These have included building and sharing knowledge; including climate change in decisions and reducing vulnerability; increasing resilience to climate change when updating, repairing, and investing in the built environment. Stormwater management systems must also adjust in order to cope with these changing conditions and increased stress.

Stormwater management facilities are important elements of the civil infrastructure that can be sensitive to climate change, particularly to precipitation extremes that generate peak runoff flows (Rosenberg et al., 2010). Inadequate stormwater management can lead to the deterioration of natural ecosystems and urban systems through hastened sediment transport, water pollutants and water quality and quantity. Urban stormwater, as stated in the Australian guidelines for urban stormwater management by ARMCANZ and ANZECC (2000), presents a management challenge in terms of: *Quantity* (i.e. flood and drainage management, stormwater reuse), *Quality* (i.e. litter, nutrients, chemicals, sediments) and *Aquatic Ecosystem Health* (i.e. aquatic habitats, riparian vegetation, stream stability and environmental flows).

According to DNRW (2008), the primary aim of an urban stormwater management system is to ensure stormwater generated from developed catchments causes minimal nuisance, danger and damage to people, property and the environment. However, these stormwater systems form part of the urban infrastructure with many different stakeholders, each of which may have different priorities. These can include not only water quality and quantity management issues, but also economic, community recreation and aesthetics aspects. Further, the impacts of climatic changes are expected to affect local and regional areas differently. Thus, the majority of adaptive actions will need to be decided and undertaken at the local and regional level.

The selection and implementation of appropriate adaptation strategies to reduce climate change impacts is a complex problem. Logically, in order to make effective use of finite resources, this adaption process will synthesise input from various stakeholders. As explained in ARMCANZ and ANZECC (2000), the difficulty stems from the lack of an appropriate single objective (criterion) for the management of all urban stormwater systems. A decision based on a single criterion reflects an oversimplification of the characteristics of the problem under consideration. A multiple objective approach is thus appropriate, considering objectives such as: *Ecosystem Health*, both aquatic and terrestrial; *Flooding and Drainage Control*; *Public health and safety*; *Economic Considerations*; *Recreational Opportunities*; *Social Considerations*; and *Aesthetic Values*.

In this context, the purpose of this research is to identify and evaluate preferred adaptation options which could reduce the vulnerability of stormwater systems to a changing climate. In order to organise and analyse the needs and desires of multiple stakeholders in a situation with multiple objectives, multiple criteria decision aid tools have been employed, specifically the analytical hierarchy process technique. This required the development of a decision making framework, based on a review of existing adaption options. Three key stakeholder groups within Gold Coast City Council, Sunshine Coast Regional Council, and Moreton Bay Regional Council were then engaged to trial the framework. The results of this case study have been utilised to make suggestions on future stormwater management, and on the refinement of the developed decision making framework.

APPROACH

Decision Making for Environmental Problems

Environmental systems are typified by uncertain behaviour and complex feedback mechanisms. When confronting anticipated changes such as climate change, there are two dilemmas for decisions maker's (DM's): *how* and *when* to adapt. The process requires consideration of many stakeholders, with different goals, and numerous adaptation alternatives. Despite this confusion and complexity, one or more responses to climate change must ultimately be chosen to facilitate the desired adaption.

Another critical question is: who should make these decisions? While climate change is a global problem that affects everyone, some people and certain communities are more affected than others, depending on their location as well as their ability to adapt. Even within communities the level of risk and impacts can vary significantly, with stormwater management being a key example of this. Therefore, the decision making process has to include all stakeholders directly or indirectly affected by changing climate. Into this mix will inevitably be disagreements about which objectives should be achieved, and which criteria or definitions should be used.

This process is a particularly delicate endeavour in democratic systems with limited resources, especially where many stakeholders have to be satisfied. The public's support of DM's efforts to reduce the impacts of changing climatic conditions depends on the dissemination of robustly debated information from experts across a range of disciplines to the community.

Clearly, a resilient and flexible adaptation alternative selection technique would be ideal for environmental decision making processes. This process would include multiple stakeholders (as DMs), who prioritise adaptation alternatives, and provide a clear identification of the sequence of implementation.

Decision making is a process of selecting from among several alternatives based on various (usually conflicting) criteria. The decision based on a single criterion would be oversimplification of the characteristics of the problem under consideration, and therefore may lead to inappropriate decisions. According to pioneers in multiple criteria decision aid (MCDA) fields, a problem can be considered as a decision problem if there are at least two criteria (in some cases conflicting) to deal with, otherwise it would be a problem of getting the right information, not a decision problem (Roy, 1968; Keeney and Raiffa, 1976).

MCDA techniques provide a powerful modelling framework for assisting complex decision-making processes, especially involving multiple criteria, goals, or objectives of a conflicting nature. These techniques concentrate on a decision analysis within a finite set of alternatives; offering tools to assist individual DMs in making decisions by eliciting and aggregating their preferences (Chen et al., 2009). The MCDA technique is used in this research because it is the most suitable approach by which to identify the priority of adaptation alternatives. Information on priority alternatives is vital in aiding DMs to design more effective adaptation options and better management plans to reduce the adverse effects of climate change.

Several multi-criteria decision aid techniques are suitable for comparing multiple criteria, simultaneously, and for providing a solution to a given problem. While there are no better or worse techniques, some techniques are better suited to a particular decision problem (Haralambopoulos and Polatidis, 2003).

The analytical hierarchy process (AHP) technique, despite some criticisms, has been selected for the current study. Criticisms of the AHP include; difficulty of conversion from verbal to numeric scale; inconsistencies imposed by the 1 to 9 scale; number of comparisons required may be large (Ramanathan and Ganesh, 1995; Macharis et al., 2004). However, AHP, owing to its flexibility to be integrated with different techniques, enables the user to extract benefits from all the combined methods and, hence, achieve the desired goal in a improved manner (Vaidya and Kumar, 2006). In addition, the AHP is set apart from other MCDA techniques because of the unique utilisation of a hierarchy structure to represent a problem in the form of a goal, criteria and alternatives (Saaty and Kearns, 1985). This allows for a breakdown of the problem into various parts for pair wise comparisons, which uses a single judgement scale. Thus, the AHP has been widely used to solve various decision problems. Examples of the recent AHP applications include: Awasthi and Chauhan (2011); Bottero et al. (2011); Crossman et al. (2011); Chen and Paydar (2012); Do et al. (2012); Gao and Hailu (2012); Sahin et al. (2013b); and Sahin and Mohamed (2013).

The Analytical Hierarchy Process (AHP)

The underlying concept of the AHP technique is to convert subjective assessments of relative importance to a set of overall scores or weights (Saaty, 1980). The AHP, based on three principles as defined by Saaty, is an Eigen value approach to the pair-wise comparison (Schmoldt, 2001). These principles are: (1) Decomposition, (2) Evaluation, and (3) Synthesis.

The AHP process starts with decomposing a decision problem into a hierarchy by listing the overall goal, criteria, and decision alternatives. In the hierarchy, the first level is the goal followed by intermediate levels corresponding to criteria and sub criteria. The lowest level contains the alternatives.

After building the hierarchy, criteria are compared in a comparison matrix, and the decision alternatives are compared in a comparison matrix with respect to each criterion. Participants are required to determine the relative priorities of each element through comparing them in a pair-wise manner, with respect to each element of immediate upper level in the hierarchy by using Saaty's nine point (i.e. 1, 2, ..., and 9) intensity scale of importance.

The relative contribution of each element on an element in the level immediately above can be determined by using the priority vector (or Eigenvector). The eigenvector method derives benefit from the information provided in the matrix, whatever the inconsistency may be. It also derives priorities based on the information without conducting arithmetic improvements on the data (Holloway, 1987). For calculating eigenvector values Saaty (1980) uses geometric mean method. That is:

$$w_i = a_i / \sum_{i=1}^{n} a_i \quad (1)$$

where: a_i ($a_1, a_2, \ldots, a_n$) is a set of eigenvector components abd w_i ($w_1, w_2, \ldots, w_n$) is a set of normalised eigenvector components.

The next step is to look for data inconsistencies. In decision analyses it is important to know the degree of decision makers' consistency, although in the AHP technique, a certain amount is desirable. Inconsistency, according to Saaty (1991), must be large enough to allow for change in our consistent understanding, but small enough to make it possible to adapt our old beliefs to new information.

The consistency index (CI) for a *nxn* reciprocal matrix proposed by Saaty is calculated from the largest eigenvalue, λ_{max}:

$$CI = (\lambda_{max} - n) / (n-1) \quad (2)$$

where, n is the number of elements being compared and $\lambda_{max} \geq n$.

The consistency index is compared to a value derived by generating random reciprocal matrices of the same size, called random consistency index (*RCI*), to give a consistency ratio (*CR*). Then, the *CR* of *nxn* matrix can be expressed as the ratio of its *CI*:

$$CR = CI/RCI \quad (3)$$

The *RCI* values are generated randomly from the scale (1/9, 1/8, 1/7,..., 1/2, 1, 2, 3,..., 9). A sample RCI table suggested by Saaty is shown below:

Number of Alternative (n)	3	4	5	6	7	8
Random Index (RCI)	0.58	0.90	1.12	1.24	1.32	1.41

If CR is less than 10% then the matrix is considered as having an acceptable consistency. In some cases, 20% CR can be tolerated but never more (Saaty et al., 1991). If the CR is not within the acceptable range then DMs are to review the problem and revise their judgments.

Finally, the local priorities are synthesised across various levels to determine the final priorities of alternatives. This can be done, as described by Saaty (1991), by multiplying local priorities by the priority of their corresponding criterion in the level above and adding them for each element in a level according to the criteria it affects.

Adaptation Issues in the Study Area

There is general consensus in the scientific community over global warming and the resultant changes in climate it will bring (Doran and Zimmerman, 2009; Anderegg et al., 2010).

In line with this global trend, Australia is already experiencing the impacts of a changing climate and is expected to face more severe extreme events (Meehl, 2007; Pearce et al., 2007; Choy et al., 2012). In particular, the Sunshine Coast Regional Council state that SEQ will see 'more intense but less overall rainfall with longer periods of drought, increases in mean sea levels over time and more extreme weather events such as cyclones and severe thunderstorms' (SCRC, 2009).

Stormwater infrastructures are considered one of the most costly and vulnerable to changing climate (Taylor, 2005). Considering this, the fact that local governments base their long-lived infrastructure investments (i.e. roads and stormwater) on design standards founded on past climate conditions suggests that for the purpose of reducing future risk (environmental, social, financial etc.), local governments need to prepare new or modified adaptation strategies (Apan et al., 2010).

This need is further exemplified by the fact that SEQ is one of the six 'vulnerability hotspots' in Australia due to its growing population and proximity to the coast (Hennessy et al., 2006).

One of the key problems in the management of stormwater is presented by the progressive changes in the Australian landscape as a result of population growth and urbanisation. Particular to the region of SEQ, population figures are around 3 million and are expected to reach as much as 4.1 million by 2026 (Queensland Office of Economic and Statistical Research, 2011).

As a result of rapid population growth and urbanisation, SEQ has seen an increase in the amount of hard surfaces (Allen Consulting Group and Australian Greenhouse Office, 2005). This coupled with the runoff issues associated with heavy precipitation after long periods of drought, poses a significant problem. Some of the impacts SEQ communities are expected to face as a result of projected climate change include:

- economic stress when required to repair and/or replace infrastructures (Homes, buildings, pipelines, roads etc.) as a result of flooding and the follow-on impacts such as land subsidence and erosion;
- public health issues resulting from flooding and drought such as physical injury, disease and emotional stress; and,
- deterioration of natural ecosystems and urban systems through hastened sediment transports, water pollutants and water quality and quantity.

Considering the community pressures brought about by a changing climate combined with non-climatic drivers such as population growth and urban development, it is evident that communities need to respond to these changes through planned adaptation and continual updating of their built and natural environments.

MODEL DEVELOPMENT

Table 1 outlines the methodology employed in establishing a hierarchical model, and displays the application of this model through analysis of stakeholder opinion drawn from three SEQ coastal cities. The AHP technique was utilised to evaluate criteria and alternatives identified through an extensive literature review and stakeholder engagements.

Table 1. Decision model development steps

	Decision Modelling Process				
Steps	Identify Stake-holders	Identify Decision Goal	Identify Decision Criteria	Identify Decision Alternatives	Evaluate Alternatives
Brainstorming	x	x	x	x	
Stakeholder engagement	x	x	x	x	x
Literature review	x		x	x	
Multi criteria decision analysis - AHP					x

Applying AHP to a given problem involves the implementation of three principles: decomposition, evaluation and synthesis. Based on these three principles the following steps were taken to implement the AHP:

1) Decomposition:
- determine the stakeholders by establishing which groups are deemed to be end-use decision makers;
- evaluate the problem and determine the goal;
- establish a list of criteria (objectives) that the stakeholders deem to be important. The use of words that cover a wide range of objectives can be useful in creating a simplified hierarchy model;
- determine a list of alternatives that are solutions to the goal. Again, the right wording can reduce the amount of alternatives to maintain simplicity; and,
- create a hierarchy model with the levels consisting of the goal up top, stakeholders underneath this, followed by the criteria and then the alternatives at the bottom.

2) Evaluation:
- make pairwise comparison of the elements within each level of the hierarchy model. This involves specifying stakeholders' preference of the different alternatives with respect to the different criteria, making judgements on what criteria each stakeholder group finds most important, and determining the weight of each stakeholder's opinion with regard to the goal.

3) Synthesis:
- calculate the CR and perform a sensitivity analysis to ensure that the data collected is suitable for use in making recommendations; and,
- use matrices to propagate level specific, local priorities to overall priorities.

Stakeholder Identification and Engagement

The goal of stakeholder identification and engagement is to establish which of the many stakeholders are most important, and which stakeholders can best represent the diverse range of needs within the community and its sectors.

A commitment to stakeholder engagement will be crucial to the success of any climate change adaption. This is reflected in a large body of literature emphasising the importance of stakeholder participation in the decision making process on environmental issues, which are complex, uncertain, and vary in time and space (van den Hove, 2000; Willows and Connell, 2003; Lim et al., 2004; Adger et al., 2007).

Thus, it is important to initially establish, with the help of the steering group or related mechanisms, who the key stakeholders are that need to be consulted. These stakeholders also become important in refining methodologies, disseminating vulnerability information, and successfully implementing adaptation strategies.

The Willows & Connell's (2003) framework stresses the importance of an open approach to decision-making, which takes into account of the legitimate interests of stakeholders and affected parties. Where appropriate, the decision process should encourage active participation from interested groups. Among other benefits, this helps to minimise the risk of overlooking potential impacts, and of failing to identify adaptation constraining decisions. It should also ensure that differences in the perception of risks and values are fully explored within the risk assessment and decision appraisal process. (Willows and Connell, 2003).

A large amount of stakeholders involved in the decision making process were identified however, including the opinions of such a large group would be unrealistic. Therefore, this study primarily targets end-use professionals that prepare stormwater management plans within local and regional governments/councils (*Engineering Services*, *Planning and Environment*, and *Finance*). These departments were engaged within Sunshine Coast Regional Council (SCRC), Moreton Bay regional Council (MBRC), and Gold Coast City Council (GCCC).

To these ends, stakeholder engagement was achieved through:

- targeted consultations with stakeholders throughout the duration of the study; and,
- conducting surveys to elicit stakeholder opinions across three regions.

Formulation of Goal, Criteria and Adaptation Options

Once the stakeholders were identified, a goal, a list of criteria (objectives) and alternatives were established on the basis of two sources:

1) Existing literature on the adaptation strategies employed currently in Australia and around the globe. This included reviewing the guidelines for urban stormwater management and design (ARMCANZ and ANZECC, 2000).
2) Consultation with stakeholder groups through workshops held by SEQCARI. This permitted for stakeholders opinion to be considered when choosing the criteria (objectives) and alternatives.

As shown in Figure 1, Section 2.4.3, the overarching goal used in the AHP structure was to adapt stormwater systems to climate change in order to minimise nuisance, danger and damage to people, property and the environment.

Adaptation options were generated from the literature review, document analyses and informed by stakeholder feedback through targeted stakeholder consultations. The review and testing of adaptation options was continued through further stakeholder engagement processes (e.g. stakeholder interviews and consultations). The adaptation alternatives identified for the AHP analyses are:

1) *Increase public awareness and participation*: Develop public education initiatives and awareness programs concerning the negative impacts of climate change on stormwater infrastructure.
2) *Monitor changing climate conditions and use of technology*: Monitor the condition of stormwater infrastructure conditions through greater use of technology to receive advance warnings, and use alternative preventive retrofitting or reconstruction to reduce the impact of changing climate.
3) *Change operational and maintenance practices*: Incorporate climate change in emergency planning by considering potential increases in frequency of flooding. Use more proactive operational and maintenance standards by redesigning or improving existing standards to withstand the new conditions due to climate change and associated extremes.
4) *Change stormwater infrastructure design standards*: Modify stormwater management infrastructure design standards to increase capacity in anticipation of increased stresses due to climate change. Many stormwater infrastructures are currently designed for the 1 in 100 year flood event. However, it is predicted that what is today's 100-year flood event from precipitation is likely to occur more frequently (i.e. every 50 or even every 20 years by the end of the current century). Therefore it is important to understand whether current design standards are sufficient to accommodate climate change.
5) *Modify Planning and land use control standards*: To prevent or reduce development in flood prone areas in anticipation of increased frequency and magnitude of flooding due to changing climate, modify land use regulations and development standards to avoid placing people and new development in vulnerable locations, and relocate existing vulnerable infrastructure and citizens to safer areas.

Finally, for this study, the following five criteria (objectives) were established for multi criteria analysis:

1) *Public health and safety*: To minimise the public risk to: public health from mosquitoes, the community/properties from flooding, and injury or loss of life.
2) *Ecosystem health*: To retain natural drainage systems, protect ecosystem health and protect existing values of waterways, wetlands, estuaries, marine and associated vegetation from development impacts.
3) *Economic considerations*: To implement stormwater management systems which are economically viable in the long-term. In many cases, the perceived cost of alternative stormwater treatment systems results in reluctance to change. However, in reality or

when demonstrated, the long-term costs may be significantly lower than conventional systems. In addition, the value of stormwater in the built environment is taken into account and reflects its true social, environmental and economic contributions.

4) *Social considerations* (aesthetic values – recreational opportunities): To ensure that community social, aesthetic and cultural values are recognised and maintained when managing stormwater. Community values now encompass concern for improved access to open space and a variety of recreation opportunities, quality of life, and aesthetic living environment, conservation of Aboriginal heritage sites, environmental protection and ecologically sustainable development.
5) *Flooding and drainage control*: To protect the built environment from flooding and waterlogging. It is important to ensure that our urban environments have minimal risk of damage from water.

Decision Hierarchy

The hierarchy structure, illustrated in Figure 1, represents the goal, stakeholders, objectives and alternatives for use with the AHP. The stakeholder groups were weighted within the hierarchy structure. This allowed the goal to be influenced by the priority given to each stakeholder groups in that level. A number of case studies regarding resource management involving multiple stakeholders are listed by Harrison and Qureshi (2000). They observed, aside from a few exceptions, that weights were not strategically assigned to stakeholder groups, or that no weights were assigned. In regards to the application of AHP for this study, it is evident that clear identification of stakeholder groups combined with assigning weights to stakeholder is deemed necessary to obtain a successful result.

The hierarchy structure shown below was utilised for all participating stakeholders as this allowed them to answer a uniform set of questions. As such, three different sets (within the same gross data population) were obtained (one for each stakeholder group) which allowed for the comparison and identification of the differences between the stakeholder groups. The use of AHP in this manner allows for flexibility when combining participant responses.

Data Collection

A survey questionnaire was developed to obtain the judgements from the nominated participants. Although there are other methods of obtaining stakeholder opinions such as group specific and non-specific focus meetings, these methods are open to intimidation and domination by individuals that result in inaccurate data.

Through sending survey questionnaires to the three nominated city councils, the data gathered is assumed to honestly represent the opinion of stakeholders without the influence of other individuals. Additionally, the ability for participants to complete the survey at their leisure should have helped ensure that judgements were well considered and not rushed.

The survey distribution process involved identifying stakeholders and their locations. Once this was done, 50 surveys were sent with the request of them being further distributed to people within the regional council. A total of twelve responses were received. Since the AHP is not a statistically based decision model, a large data group is not inherently necessary. The AHP functions on the geometric mean of the comparison ratings opposed to individual ratings (Duke and Aull-Hyde, 2002).

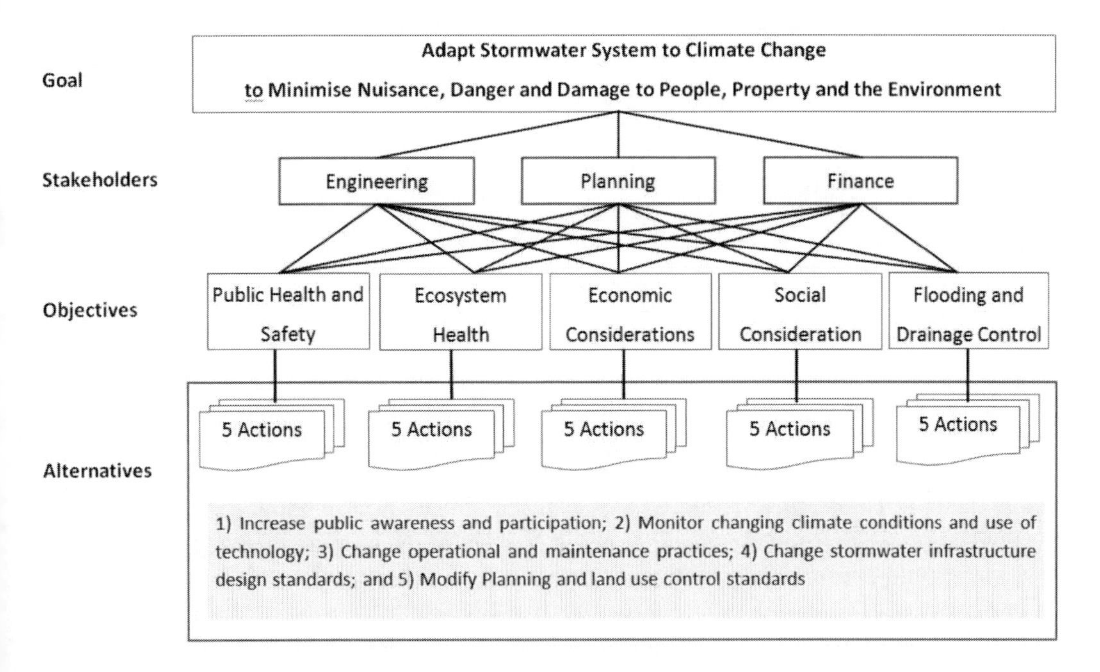

Figure 1. AHP model for adapting stormwater systems to climate change.

RESULTS

In order to elicit a priority ranking of the adaptation alternatives, the stakeholder participants were asked to make pairwise comparisons on the adaption alternatives with respect to the goal of 'Adapting Stormwater Systems to Climate Change'.

On a practical level; at each stage of the hierarchy, participants were required to indicate which of two options were more important, or more influential than another. For example, at the stakeholder level "What objective is more important to the Planning Department?", or at the criteria level "Which action has more impact on Ecosystem Health?". In this way each participant was questioned on what was important to their own department, and what was important to other departments. Then, utilising participant judgements from the three stakeholder groups within the three SEQ regional councils, the priority of each alternative was evaluated and interpretations drawn from these results.

Consistency Ratio (CR)

Before interpretation and recommendations were provided, the results were checked for consistency to confirm their validity. Consistency is defined as holding an unchanging opinion over the period of an activity. Although the AHP does not demand perfect consistency to elicit a rank of alternatives, there is a threshold of CR<10%; for environmental problems that pose high variability Saaty suggest that CR<20% is acceptable (Saaty and Kearns, 1985; Saaty et al., 1991). The survey results showed a CR for SCRC of 5.4%, MBRC of 1.4%, and for GCCC of 3.2%; well within the acceptable level.

DATA ANALYSIS

Prioritisation of Stakeholders

Combining the judgements made by all participants showed that the most important stakeholder department in the decision making process was *Planning.* The second and third most important stakeholder departments were respectively considered to be *Engineering* and *Finance*. This ranking priority was however not displayed when using the judgements made by separate stakeholder groups. It was found that in all cases each stakeholder department did not assign themselves with the most importance and in the case of the *Finance* department, they assigned themselves to be by far the least importance. This could infer that each stakeholder does not necessarily hold themselves with the highest importance. Furthermore, it could be concluded that each stakeholder group is either of the opinion that they don't have as much input as they actually do, or they do not want the greatest responsibility in the decision making process. Considering this observation, it is important to structure a hierarchy model that will use a combined participant judgement to assign the weight of each stakeholder group (see Section 3.2.2).

Prioritisation of Criteria

When combining the judgements made by all participants, it was found that the most important criteria to consider when comparing the five adaptation alternatives is *Public Health and Safety* closely followed by *Economic Considerations* (see Figure 2).

This priority ranking however was not reflected by any of the individual stakeholder groups which further exemplify the diverse opinions and the need for MCDA (see Figure 3). The least preferred criterion was *Social Considerations* which is a general trend for all three stakeholder groups.

It was expected that *Public Health and Safety* would gain the highest priority due to the endangerments recently witnessed in the SEQ floods. It was also expected that *Economic Considerations* would closely follow due to the participant's knowledge of the economic pressure associated with floods combined with the current sluggish economy. It is interesting that according to the opinion of all participants, both the *Finance* and *Planning Departments* favour *Economic Considerations* over *Public Health and Safety*.

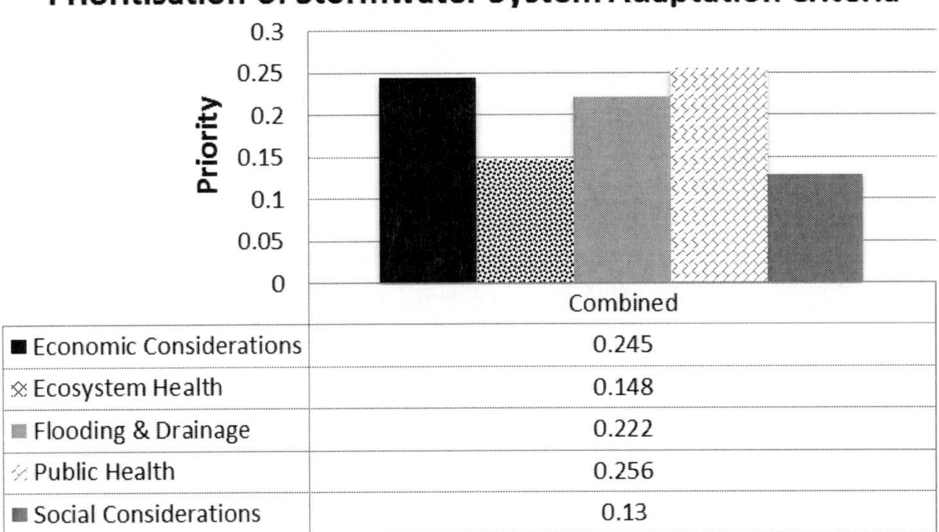

Figure 2. Combined criteria priorities from all stakeholders.

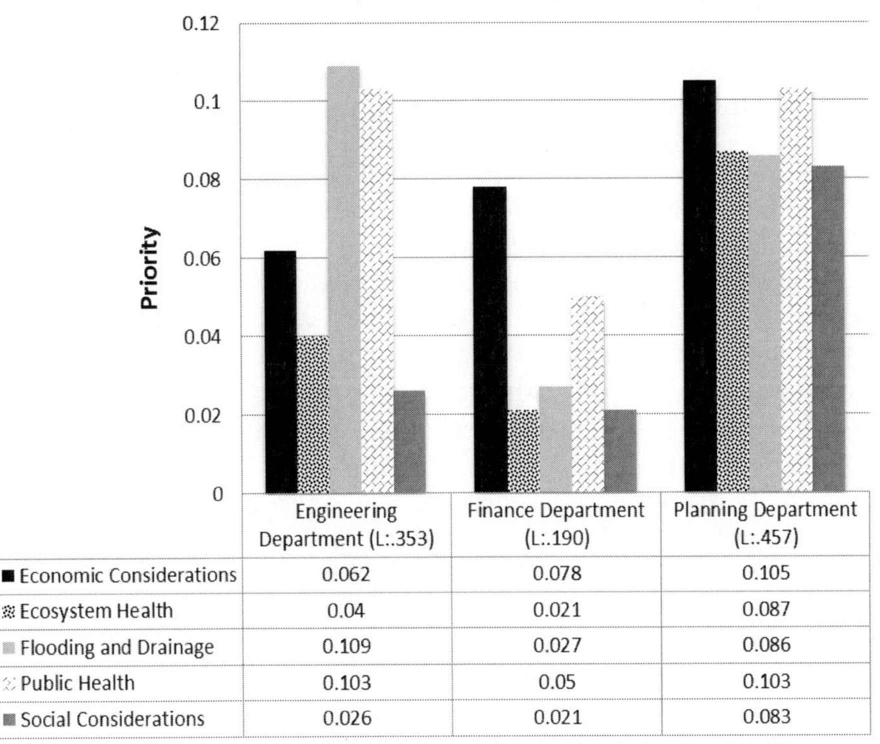

Figure 3. Criteria priorities with respect to each stakeholder group; according to the judgement made by all participants.

This could suggest that stakeholders believe themselves, or their colleagues, to be influenced by their professional context. It could also relate to the recent media hype surrounding the two speed economy and/or both departments not considering the issue of climate change and stormwater system vulnerability to be of great concern. DM's must take into account the current social conditions and consequent media hype that may affect the participant's judgment.

A noticeable trend can be seen when comparing the two lowest preferred criteria; viz. *Ecosystem Health* and *Social Considerations*. It was observed that for both the combined as well as the individual stakeholder opinions, *Ecosystem Health* was considered more important than *Social Consideration*; although by a small margin. This suggests that participants are well aware of the environmental threats associated with inadequate stormwater systems such as flooding and droughts, and consider social considerations such as aesthetic and recreational values of lesser importance. Whether the general public would hold the same opinion is unknown.

The aforementioned priorities were elicited through using the judgements made by all participants. It was found that using the judgements made by each stakeholder group separately, resulted in a priority ranking that differed from the ranking expressed in Figure 3. This suggests that allowing stakeholders to make judgements on what they perceive to be the opinion of another stakeholder group is inaccurate. Similarly, it was found that when using the judgements made by each city council separately, the priority ranking differed from the ranking presented in Figure 2. This proposes that each city council has differing opinion based on their regional conditions (i.e. funding, adaptation capacity etc.).

Prioritisation of Alternatives

Using combined stakeholder judgement, it was found that the most preferred adaptation alternative was *Modify Planning and Land Use Control Standards*, followed by *Change Stormwater Infrastructure Design Standards* (see Figure 4).

These main concerns are reflected in the individual stakeholder priorities which suggest a strong consensus on the prioritisation of these two alternatives. The reasons behind the stakeholder's first and second preferences are numerous.

One such reason being that due to the extensive infrastructure damage as a result of the recent SEQ floods, it has become apparent that housing and infrastructure is built too close to natural flow paths.

The least preferred alternative was *Monitor Changing Climate Conditions and Use of Technology,* which again was a general trend across all three stakeholder groups. This result seems rather unusual due to the fact that there is a considerable amount of uncertainty when it comes to local-scale climate projections.

Considering this uncertainty, its unexpected that the *Finance Department* also expressed this priority, especially when there are huge costs associated with implementing climate adaptation strategies. This could infer that end-use professionals entirely trust the climate projections and are willing to bare the upfront costs in the chance that they are correct. It is also possible that the participants involved did not fully understand this alternative and therefore neglected giving it priority.

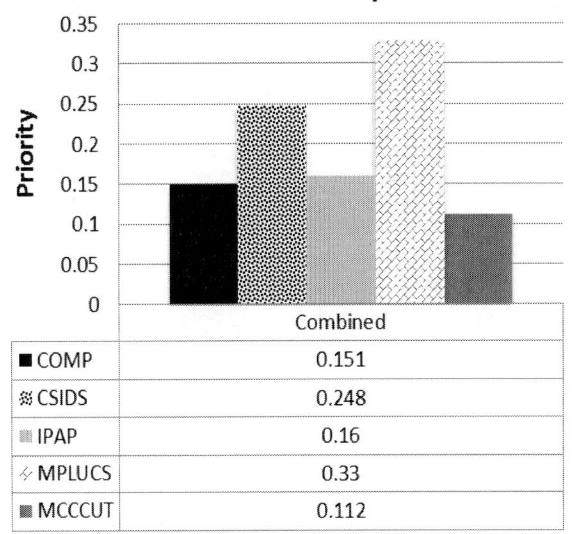

COMP: Change Operational and Maintenance Practices; CSIDS: Change Stormwater Infrastructure Design Standards; IPAP; Increase Public Awareness and Participation; MPLUCS; Modify Planning and Land Use Control Standards; MCCCUT: Monitor Changing Climate Conditions and Use of Technology

Figure 4. Combined adaption alternative priorities for all stakeholder groups.

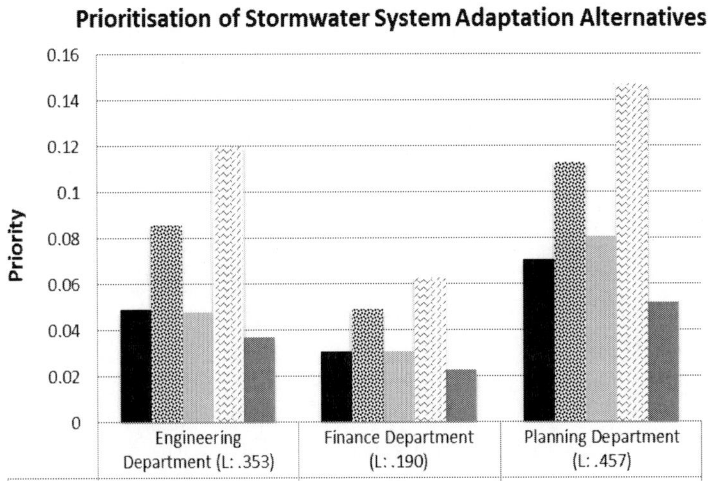

Prioritisation of Stormwater System Adaptation Alternatives

Priority

0.16, 0.14, 0.12, 0.1, 0.08, 0.06, 0.04, 0.02, 0

	Engineering Department (L: .353)	Finance Department (L: .190)	Planning Department (L: .457)
COMP	0.049	0.031	0.071
CSIDS	0.086	0.049	0.113
IPAP	0.048	0.031	0.081
MPLUCS	0.12	0.063	0.147
MCCCUT	0.037	0.023	0.052

COMP: Change Operational and Maintenance Practices; CSIDS: Change Stormwater Infrastructure Design Standards; IPAP; Increase Public Awareness and Participation; MPLUCS; Modify Planning and Land Use Control Standards; MCCCUT: Monitor Changing Climate Conditions and Use of Technology

Stakeholder Departments & Adaptation Alternatives

Figure 5. Adaption priorities with respect to each stakeholder group; according to the judgement made by all participants.

From Figure 5 it can be seen that there is a distinct link between *Increase Public Awareness and Participation* and *Change Operational and Maintenance Practices*, with both the individual and aggregated rankings holding similar priority. This is fitting when it is considered that increasing public awareness of stormwater system vulnerability to extreme events (e.g. blocked drains) can supplement a council's maintenance practices. Another interesting result was that the finance department held the highest priority for the two most expensive options. This suggests that regardless of professional role there is a consensus that physical adaption must occur to mitigate future risks.

Similar to the criteria results it was found that considering the adaptation alternative judgements of each stakeholder group individually resulted in priority rankings that differed from the combined rankings illustrated in Figure 4. This again suggests that stakeholders cannot accurately predict what the opinion of a stakeholder in a different professional context may have.

It was also found that when using the judgements made by each city council separately, the priority ranking differed from the ranking presented in Figure 3. Thus regional conditions and organisational culture could be argued to impact upon DM.

CONCLUSION

This study adopted a multiple-criteria decision approach and engaged stakeholders to evaluate alternatives in adapting stormwater systems to climate change, within the local governments of three SEQ cities. These stakeholders completed a survey composed under AHP, allowing the results to be expressed numerically. It was observed that the prioritisation of adaptation criteria and alternatives varied noticeably between using the judgement of combined participants compared to the judgements of each stakeholder groups. This suggests that although there is an acceptable level of consistency within the results used, there exists some disparity between the stakeholders opinion on:

1) the importance or ranking of each stakeholder in the decision making process; and,
2) what each stakeholder group believes to be the opinion or preference of other stakeholder groups.

When looking at the ranking of alternatives according to the judgments made by each stakeholder group, it was interesting to observe that in most cases the priorities expressed by each stakeholder were also assumed to be the priority for the other two stakeholder groups; although the strengths of preference were not the same. This clearly demonstrates that allowing stakeholders to make judgements on what they perceive to be the opinion of another stakeholder group is inappropriate.

Another observed trend was the difference between the weights each stakeholder assigned themselves in comparison to the weight collectively assigned by the 'Combined Participant Judgement'. This could infer that each stakeholder does not necessarily hold themselves with the highest importance. In the case of the *Finance Department*, they were found to hold themselves with the lowest opinion weight. Furthermore, it could be concluded that each stakeholder group either believes that they don't have as much input as they actually do, or that they do not want the greatest responsibility in the decision making process. Considering this observation, it is important to structure a hierarchy model that will utilise a combined participant judgment to assign the weight of each stakeholder group.

Finally, it was noted that in all cases (combined, individual stakeholder groups, and regional councils) there was a clear distinction between the significantly higher prioritised alternatives and the remaining lower and less preferred options. Suggesting there's a clearly identified preference of two alternatives: viz. *Modify Planning and Land Use Control Standards* and *Change Stormwater Infrastructure Design Standards.* The remaining three alternative were clustered together (similar priorities) and separated significantly from the two most preferred. This may imply that, although there are two clearly preferred alternatives, the participants are unsure as to the implementation of the remaining alternatives. In the case of the most preferred options being unviable or prohibitively expensive, this cluster could pose some negative consequences such as implementation delay.

In summary, the MCDA AHP model has been effective in identifying the opinions of individual stakeholders, and synthesising these into a combined judgement, revealing a preference for adaption that would not have otherwise been transparent. The results indicate that local government professionals in a range of departments believe fundamental change to planning, land use and design standards is required to mitigate the risks of climate change for stormwater systems in SEQ.

REFERENCES

Adger, W. N., et al., 2007. Assessment of adaptation practices, options, constraints and capacity. *Climate Change 2007: Impacts, Adaptation and Vulnerability. Contribution of WG-II to the 4AR Assessment Report of the IPCC.* Cambridge, UK.

Allen Consulting Group and Australian Greenhouse Office, 2005. *Climate change risk and vulnerability : promoting an efficient adaptation response in Australia : final report,* Canberra, A.C.T., Australian Greenhouse Office.

Anderegg, W. R. L., Prall, J. W., Harold, J. & Schneider, S. H. 2010. Expert credibility in climate change. *Proceedings of the National Academy of Sciences,* 10712107-12109.

Apan, A., et al., 2010. The 2008 floods in Queensland: a case study of vulnerability, resilience and adaptive capacity. Report for the National Climate Change Adaptation Research Facility. Gold Coast, Australia: University of Southern Queensland.

ARMCANZ and ANZECC, 2000. *Australian guidelines for urban stormwater management, 2000,* Canberra, A.C.T., Agriculture and Resource Management Council of Australia and New Zealand and Australian and New Zealand Environment and Conservation Council.

Awasthi, A. and Chauhan, S. S., 2011. Using AHP and Dempster–Shafer theory for evaluating sustainable transport solutions. *Environmental Modelling & Software,* 26(6), 787-796.

Bottero, M., Comino, E. and Riggio, V., 2011. Application of the Analytic Hierarchy Process and the Analytic Network Process for the assessment of different wastewater treatment systems. *Environmental Modelling & Software,* 26(10), 1211-1224.

Chen, Y., et al., 2009. A DEA-TOPSIS method for multiple criteria decision analysis in emergency management. *Journal of Systems Science and Systems Engineering,* 18(4), 489-507.

Chen, Y. and Paydar, Z., 2012. Evaluation of potential irrigation expansion using a spatial fuzzy multi-criteria decision framework. *Environmental Modelling & Software,* 38(0), 147-157.

Choy, D. L., et al., 2012. Adaptation Options for Human Settlements in South East Queensland – Main Report, unpublished report for the South East Queensland Climate Adaptation Research Initiative. Brisbane, Australia: Griffith University.

Crossman, N. D., Bryan, B. A. and King, D., 2011. Contribution of site assessment toward prioritising investment in natural capital. *Environmental Modelling & Software,* 26(1), 30-37.

DNRW, 2008. Queensland urban drainage manual. Second ed. Brisbane, Qld.: Department of Natural Resources and Water.

Do, H. T., et al., 2012. Design of sampling locations for mountainous river monitoring. *Environmental Modelling & Software,* 27–28(0), 62-70.

Doran, P. T. & Zimmerman, M. K. 2009. Examining the scientific consensus on climate change. *EOS,* 90 (3)22-23.

Duke, J. M. and Aull-Hyde, R., 2002. Identifying public preferences for land preservation using the analytic hierarchy process. *Ecological Economics,* 42(1-2), 131-145.

Gao, L. and Hailu, A., 2012. Ranking management strategies with complex outcomes: An AHP-fuzzy evaluation of recreational fishing using an integrated agent-based model of a coral reef ecosystem. *Environmental Modelling & Software,* 31(0), 3-18.

Haralambopoulos, D. A. and Polatidis, H., 2003. Renewable energy projects: structuring a multi-criteria group decision-making framework. *Renewable Energy,* 28(6), 961-973.

Harrison, S. R. and Qureshi, M. E., 2000. Choice of stakeholder groups and members in multicriteria decision models. *Natural Resources Forum,* 24(1), 11-19.

Hennessy, K., Macadam, I. and Whetton, P., 2006. Climate Change Scenarios for Initial Assessment of Risk in Accordance With Risk Management Guidance. Aspendale, Vic., Australia: CSIRO Marine and Atmospheric Research.

Holloway, C., 1987. Analytical Planning - the Organization of Systems - Saaty,Tl, Kearns,Kp. *Long Range Planning,* 20(1), 144-145.

Keeney, R. L. and Raiffa, H., 1976. *Decisions with multiple objectives : preferences and value tradeoffs,* New York, Wiley.

Lim, B., et al., 2004. *Adaptation Policy Frameworks for Climate Change,* Cambridge, Cambridge University Press.

Macharis, C., et al., 2004. PROMETHEE and AHP: The design of operational synergies in multicriteria analysis. Strengthening PROMETHEE with ideas of AHP. *European Journal of Operational Research,* 153(2), 307-317.

Meehl, G. A., 2007. IPCC Climate Change 2007: The Physical Science Basis: Contribution of Working Group I to the Fourth Assessment Report of the Intergovernmental Panel on Climate Change. Nature Publishing Group.

Pearce, K., et al., 2007. Climate change in Australia Technical Report 2007.

Queensland Office of Economic and Statistical Research, 2011. *Local government areas : Queensland Government population projections to 2031,* City East, Qld., Office of Economic and Statistical Research.

Ramanathan, R. and Ganesh, L. S., 1995. Energy resource allocation incorporating qualitative and quantitative criteria: An integrated model using goal programming and AHP. *Socio-Economic Planning Sciences,* 29(3), 197-218.

Rosenberg, E. A., et al., 2010. Precipitation extremes and the impacts of climate change on stormwater infrastructure in Washington State. *Climatic Change,* 102(1-2), 319-349.

Roy, B., 1968. Ranking and Choice in Pace of Multiple Points of View (Electre Method). *Revue Francaise D Informatique De Recherche Operationnelle,* 2(8), 57- 75.

Saaty, T. L., 1980. *The Analytic Hierarchy Process,* New York, NY, McGraw-Hill.

Saaty, T. L. and Kearns, K. P., 1985. *Analytical Planning : The Organization of Systems* Oxford, New york, Pergamon Press.

Saaty, T. L., Kearns, K. P. and Vargas, L. G., 1991. *Analytical planning - the organization of systems* Pittsburg, PA, RWS Publications.

Sahin, O. and Mohamed, S., 2009. Decision Dilemmas for Adaptation to Sea Level Rise: How to, when to? *In:* SUN, H., JIAO, R. & XIE, M. (eds.) *IEEE 2009 International Conference on Industrial Engineering and Engineering Management.* Hong Kong: IEEE.

Sahin, O., Stewart, R. A. and Helfer, F., 2013. Bridging the Water Supply-Demand Gap in Australia: A Desalination Case Study. *European Water Resources Association (EWRA) 8th International Conference.* Porto, Portugal: European Water Resources Association.

Sahin, O. and Mohamed, S., 2013. A spatial temporal decision framework for adaptation to sea level rise. *Environmental Modelling & Software*, 46(0), 129-141.

Sahin, O., et al., 2013b. Assessment of Sea Level Rise Adaptation Options: Multiple-Criteria Decision-Making Approach Involving Stakeholders. *Structural Survey,* 31(4)(4).

Schmoldt, D. L., 2001. *The analytic hierarchy process in natural resource and environmental decision making,* Dordrecht ; Boston, Kluwer Academic Publishers.

Siems, R., Sahin, O., Talebpour, R., Stewart, R. & Hopewell, M. 2013. Energy intensity of decentralised water supply systems utilised in addressing water shortages. *European Water Resources Association (EWRA) 8th International Conference.* Porto, Portugal: European Water Resources Association.

Taylor, A., 2005. Guidelines for Evaluating the Financial, Ecological and Social Aspects of Urban Stormwater Management Measures to Improve Waterway Health. *Technical Report.* Victoria, Australia: The Cooperative Research Centre for Catchment Hydrology, Monash University.

Vaidya, O. S. and Kumar, S., 2006. Analytic hierarchy process: An overview of applications. *European Journal of Operational Research,* 169 1-29.

van den Hove, S., 2000. Participatory approaches to environmental policy-making: the European Commission Climate Policy Process as a case study. *Ecological Economics,* 33(3), 457-472.

Willows, R. I. and Connell, R. K., 2003. Climate adaptation: Risk, uncertainty and decision-making. *UKCIP Technical Report.* Oxford, UK: UKCIP.

In: Water Conservation
Editor: Monzur A. Imteaz

ISBN: 978-1-62808-993-6

Chapter 5

PUBLIC ACCEPTANCE OF ALTERNATIVE WATER SOURCES

Ana Kelly Marinoski*[*]*, Arthur Santos Silva, Abel Silva Vieira and Enedir Ghisi

Federal University of Santa Catarina,
Department of Civil Engineering, Laboratory of Energy Efficiency in Buildings, Florianópolis-SC, Brazil

ABSTRACT

Water scarcity due to population growth and global climate change is a driving force to modify the way water is managed. Alternative water sources such as rainwater and grey water are a solution to ease water shortages. Despite the potential to supply residential buildings with alternative water sources, the doubt whether these sources can be accepted by the general public remains. The present chapter addresses this issue by performing interviews in 44 low-class houses in Southern Brazil. Moreover, water end-uses and the water consumption pattern for different plumbing fixtures and appliances were also estimated by assessing 11 and 48 low-class households, respectively. Water end-uses were estimated by means of interviews and measurements or estimates of the water flow of each plumbing fixture. The acceptance data of rainwater and grey water usage were correlated with the number of occupants in each house, the total and per capita income, age and formal education level using the Pearson's correlation and p-value with 90% confidence. Results show that the shower has the greatest water end-use with 32.7% of representativeness, followed by the toilet flushing with 19.4% and the kitchen tap with 18.0%. The minimum average water consumption was 153 litres per occupant per day. The acceptance results showed that from 27 to 59% of the householders would use rainwater for potable end-uses, and from 84 to 96% would use it for non-potable end-uses. Toilet flushing had the greatest acceptance for rainwater usage. From 80 to 91% of the householders would use grey water from their own houses for non-potable end-uses, and from 73 to 82% would use it originated from other houses. In an attempt to understand these results, the correlation statistics showed that the older people are, the less they accept grey water as an alternative water source; the greater the degree of

[*] Corresponding Author address: Email: anakmarinoski@gmail.com.

instruction, the less the acceptance of rainwater; the greater the number of occupants in the house, the less the acceptance of rainwater for washing dishes. The understanding of people's approval of alternative water sources is crucial so as to develop effective strategies to overcome water scarcity in the urban environment.

Keywords: Acceptance of alternative water sources, public perception, water end-uses, rainwater and grey water usage, low-class houses

INTRODUCTION

This chapter aims at understanding the acceptance of alternative water sources (i.e. rainwater and grey water) in residential buildings in order to reduce potable water consumption in the residential sector. Therefore, analyses regarding the acceptance of alternative water sources, daily water end-uses and consumption patterns in low-class houses were assessed in the metropolitan region of Florianópolis, Southern Brazil.

Different samples of low-class houses were used for each analysis. For the analyses of the acceptance of alternative water sources, water end-uses, and water consumption patterns, 44, 48 and 11 households were assessed, respectively. All houses evaluated in this chapter are located in the same region. Sample sizes were different in each analysis due to both the willingness of householders to participate in different phases of the study, and the quality of the data generated. However, some households are repeated within the samples. Out of the 48 houses within the water end-uses sample, 24 and 9 are also included in the acceptance of alternative sources sample and the water consumption pattern sample, respectively.

BACKGROUND

Alternative water sources have been used in order to decrease potable water demand. In turn, this promotes a reduction in the abstraction of water from rivers, lakes, reservoirs, and groundwater (Gonçalves and Silva, 2012). The use of grey water may also promote a reduction in wastewater streams collected and treated by wastewater utilities, and a reduction of the total energy consumption associated with wastewater services on an urban scale (Vieira, 2012).

Several stakeholders in the water management process (i.e. governmental agencies, private sector and consumers) are increasingly aware of the need to implement water supply strategies which encompass the use of alternative water sources in order to meet future water demands in cities (Morales-Pinzón et al., 2012). Notwithstanding the water conservation benefits promoted by alternative water sources, their use depends on several factors, including: technical, economic, and environmental feasibility aspects, water quality requirements, and public acceptance (Santos and Froehner, 2007). Among these factors, public acceptance is a key parameter to the success of water strategies in which alternative water sources are used (Mancuso, 2003). Despite the economic and environmental feasibility to use alternative water sources, without public acceptance they cannot be implemented.

ACCEPTANCE OF ALTERNATIVE WATER SOURCES WORLDWIDE

A number of studies addressing the public acceptance of alternative water sources have been carried out worldwide.

In Australia, Ryan et al. (2009) studied socioeconomic and psychological indicators regarding the use of alternative water sources for irrigation in the region of Canberra. The study was performed through an internet survey with 354 households. The results indicated that there is a relationship between the use of alternative water sources (i.e. grey water and rainwater) with socioeconomic (i.e. age, gender, income and education) and psychological variables. Households that used rainwater had a greater understanding of alternative water source strategies. More specifically, female participants and low-income residents accepted more the use of grey water for irrigation; participants who used grey water for irrigation accepted more the use of rainwater as well as water reclamation schemes. Furthermore, it was found that most of the participants think that future alternative water strategies ought to consider the concomitant use of different alternative water sources in a holistic approach. Finally, the research has indicated that the potable water demand of the households could be substantially reduced by installing large-scale systems for grey water reclamation and/or rainwater harvesting.

Muthukumaran et al. (2011) conducted a survey to gain insight into the public acceptance of grey water supply in Australia. The results revealed a high acceptance of grey water for uses such as toilet flushing and irrigation. However, it was found that the acceptance significantly decreased for end-uses with direct contact with water, such as water for cleaning.

Hurlimann and McKay (2007) found the acceptance of reclaimed water for non-potable household in an urban community in South Australia. They assessed the acceptance of householders for grey water considering different parameters (i.e. colour, odour, salinity levels and price) and end-uses (irrigation, washing machine, and toilet flushing). This study aimed to spell out the public willingness to pay for the treatment of different grey water quality levels. The results indicate that the most desirable grey water characteristics are: low salinity for irrigation, low colour for washing machine, and low costs for toilet flushing. The respondents would be more willing to pay for an increase in the quality of reclaimed water for washing machine.

Also in Australia, Po et al. (2003) investigated the reasons which underpin the large theoretical acceptance of grey water, but the large refusal to implement it. According to the authors, the main factors responsible for such refusal are: (i) the sense of disgust and yuck factor; (ii) possible health risks; and (iii) cost. Pursuant to Po et al. (2003), the disgust in using grey water is possibly due to both particles in the water, and possible health risk.

Mankad (2012) undertook a literature review about the public acceptance of alternative water sources from decentralized water supply systems in urban regions in Australia. Recommendations were made to promote the adoption of decentralized systems, encouraging researchers to address issues related to different public concerns. This, in turn, may influence the decision-making process towards a more widespread use of alternative water sources.

In another study, Mankad and Tapsuwan (2011) conducted an analysis of key social and economic factors which influence the acceptance and adoption of decentralized water systems. They carried out a literature review focused on the social aspects related to alternative sources of water for domestic use. The researchers reported that most communities were open to alternative water sources for domestic use; however, the acceptance was highly dependent on the level of contact with the water. It was found that acceptance and adoption of alternative water sources were mainly influenced by risk perception and the local culture.

Other studies conducted in Australia also addressed the public perception and acceptance of alternative water sources. Hurlimann et al. (2009), Dolnicar et al. (2010), Dolnicar et al. (2011) evaluated the acceptance of reclaimed water and desalination facilities; Hurlimann and Dolnicar (2010) evaluated the acceptance of indirect potable reuse. Mankad et al. (2013) analysed the public acceptance of rainwater in residential buildings. Hurlimann and Dolnicar (2012) examined how the media interferes in the knowledge and acceptance of Australians about alternative water sources.

In the United States, Boyer et al. (2012) quantified the risks, benefits and impacts of alternative water sources regarding the perception of decision makers to support the adoption of different alternative water sources involved in planning for the urban water supply. The risks and benefits were categorized as ecological, economical and human health impacts. The results showed that, for the evaluated alternative water sources, the social acceptance occurs according to the level of impact. Ecological impacts had the greatest influence on the decision makers choices.

In Spain, Domènech and Saurí (2010) found four factors (Figure 1) which influence the public acceptance for grey water reclamation schemes, including: possible health risks, operation regimes, cost, and environmental awareness. In addition, the authors emphasize that improving the level of knowledge of users about system operation could reduce the risk of social rejection of grey water reclamation schemes.

In Syria, Mourad et al. (2011) analysed the public acceptance to use grey water for toilet flushing in the city of Sweida. In this region, households are adapted to water scarcity, as potable water is available only once a week. Therefore, each household, with five people on average, usually has one reservoir with minimal capacity of 2 m³ to meet the weekly water demand. The survey showed that 83% of respondents supported the reuse of treated grey water for toilet flushing or irrigation. The other 17% were not aware of safe treatment systems for grey water, and then did not support grey water reclamation. About 10% of respondents were already using untreated grey water from laundry for irrigation or cleaning.

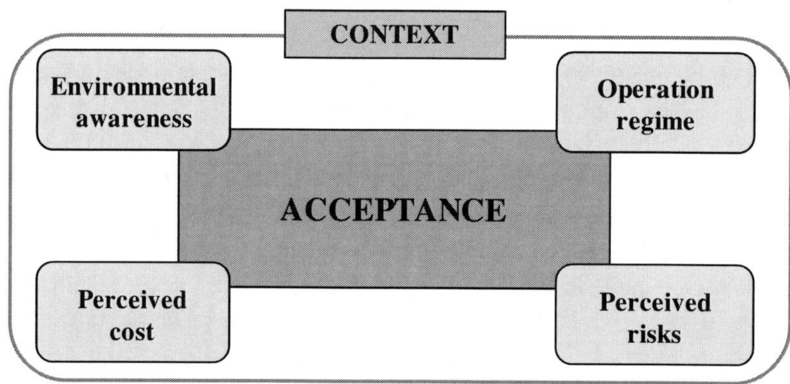

Source: Domènech e Saurí (2010).

Figure 1. Factors influencing public acceptance of grey water reuse.

In Israel, Friedler et al. (2006) conducted a survey with 256 people so as to determine the attitude of the Israeli public about various options for urban water reclamation. The survey included 21 reclamation options, which were grouped into three categories: low, medium and high contact. The results showed that the majority of interviewees supported the use of reclaimed water for landscape irrigation (95%), toilet flushing (85%) and fire fighting systems (96%). In options with high contact, such as washing clothes and aquifer recharge for drinking purposes, there was less support with 38 and 11% of support, respectively. For low level of contact option, the support was lower than expected, including a support of 86% for irrigation of vegetables, 62% for recharging aquifers for agricultural irrigation, and 49% for irrigation of fruit plantations. Water saving potential was the main factor underpinning the support for reclaimed water usage among respondents. Results also indicated that middle-aged adults in Israel are more likely to accept reclaimed water usage than young adults.

In Oman, Jamrah et al. (2004) undertook a survey addressing the public acceptance of treated grey water usage. Among the respondents, 84% were in favour of grey water usage; 74% believed it would bring financial benefits; and 82%, 72% and 42% thought it would suit irrigation, toilet flushing, and car washing purposes, respectively. However, 61% and 47% of the interviewees believed that the reclamation of grey water would be associated with environmental degradation and health risks, respectively. Surprisingly, 16% were favourable to the use of treated grey water for drinking purposes.

In Brazil, Dias et al. (2007) evaluated the level of knowledge and acceptance of the population in relation to residential rainwater harvesting systems in the city of João Pessoa, Paraíba State. In this study, 800 households were interviewed in order to have a representative sample of the studied population. The results showed that 66.1% of the interviewed population are aware of rainwater usage, and 54.4% have already used it. The study showed no correlation between the awareness about rainwater use and the level of formal education of interviewees. Among respondents who never attended school, 79.4% were aware of techniques to use rainwater. This percentage is similar to values for the group of people who had incomplete tertiary education (84.4%). This study also reviewed that 45.6% of people who were aware of rainwater usage in houses did not use it due to varying reasons, including: (i) lack of financial incentives for 70.4% of them (21.2% of the total population); (ii) technical difficulty for 47.1% of them (14.2% of total population); (iii) lack of interest for 6.2% of them (1.9% total population); and (iv) health risk concerns for 1.7% of them (0.5% in total population).

Also in Brazil, Martinetti et al. (2007) analysed the perception of people from different social classes regarding 19 alternatives for wastewater treatment. Results indicate that the selection of wastewater treatment systems is mainly based on capital costs, system footprint, number of households served, complexity of operation and maintenance.

Garcia (2011) investigated how socioeconomic characteristics influence on water consumption patterns in low-class households in Bahia, Brazil. In total, 147 households located in a low-class region in Salvador were studied. Results indicate that the average water consumption among households was equal to 10.6 m^3 per month, in which 55% consumed up to 10 m^3 per month. The consumption per capita was around 101 litres per day. Socioeconomic characteristics were directly correlated with water consumption patterns in households, including: number of residents, number of bedrooms and percentage of the household income spent with water services. Among interviewees, 59% were not aware of their water consumption patterns. For most of the households (54%), financial incentives are the most important aspects that water efficient strategies must promote. Finally, the author concluded that the tariff system used by the local water utility (i.e. fixed rates for water consumption up to 10 m³) may jeopardize the adoption of water efficient measures in low-class houses in the region; despite the importance of financial incentives for the low-class population, no financial benefits can be achieved due to the fixed water rate.

Also in Bahia, Almeida (2007) undertook a survey through questionnaires applied to 379 households located in the city of Feira de Santana. The study aimed at characterizing the water consumption patterns and the public acceptance to use reclaimed water and rainwater. Among the respondents, 90% were in favour of rainwater and grey water use as alternative water sources, and 65% showed some knowledge about the use of such water sources. In accordance to respondents, this knowledge was acquired from television for 53% of them, and from friends and family for 32% of them. Different aspects were considered the most important outcome from alternative water use, in which: (i) 34.7% of the respondents affirmed that water security is the most important aspect; (ii) 31.4%, financial benefits; and (iii) 18.5%, conservation of natural resources. Probably, water scarcity was considered an important aspect as Feira de Santana is located in a semi-arid region.

Among interviewees, 74.5% would use both alternative water sources in their residences; 10.3% would use only rainwater; 7.1% would use only grey water; and 7.1% would use neither. Figure 2 illustrates the level of acceptance to use treated grey water for different residential water end-uses.

It was observed that the population is willing to save, not necessarily for the conservation of natural resources, but mainly because of water scarcity in the region and saving resources. Thus, we note that the results obtained by Almeida (2007) differ from those found by Garcia (2011) mainly because the region studied by the first author lies in the semi-arid Brazilian, and therefore water scarcity was more relevant than the financial issue.

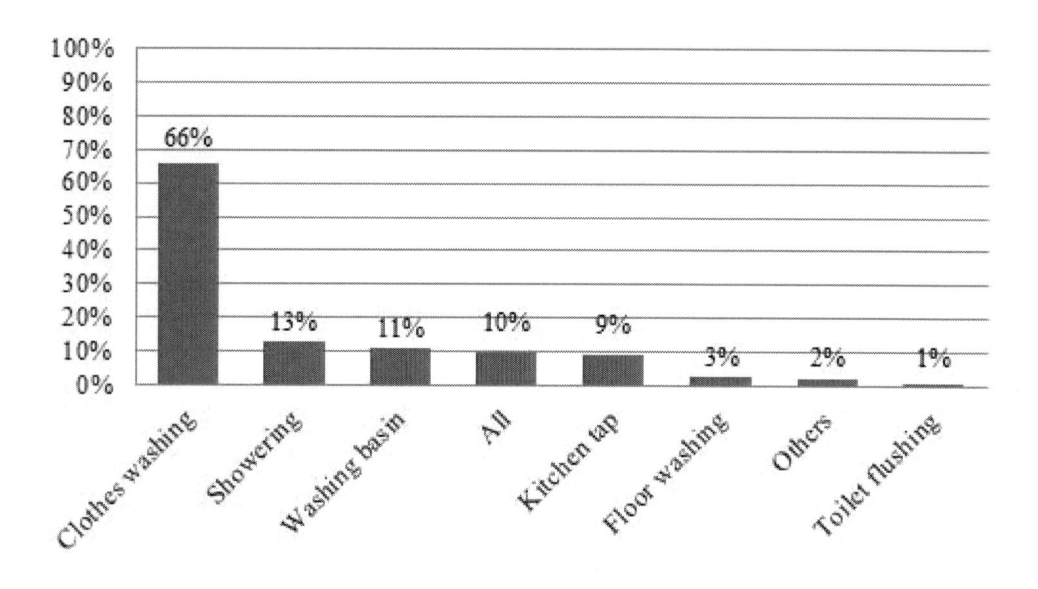

Source: Based on Almeida (2007).

Figure 2. Treated effluents from different residential water end-uses that would be acceptable for water reuse in Feira de Santana, Bahia.

In the study of Almeida (2007), the income of 70% of the households was up to five Brazilian minimum wage (i.e. R$ 403.00 and US$ 204.65 in 2007). Most of the houses were single storey with three bedrooms, one bathroom and one kitchen. The average age of respondents was 32 years, which indicates that the study group is composed mostly by young adults. According to the author, the younger people are, the more prone they are to accept changes. Hence, it is likely that young adults accept more new water management approaches than older people. Among respondents, 29.7% had primary education, 13.4% had incomplete secondary education, 31.1% had complete secondary education, and 2.8% had tertiary education. For the authors, the lack of education can influence directly the awareness of people about both their societal rights and duties, and the environmental impacts of their habits. Cohim and Cohim (2007) studied the acceptance to use reclaimed water in residential buildings in Belo Horizonte, Brazil. The authors found that the acceptance increases when the water is reclaimed within the buildings, and decreases when it is from larger systems at suburbs and communal scales.

Hafner (2007) also states that the public acceptance to use grey water depends on the exposure of users to the end-use it is intended to. In addition, the authors suggest that higher education and income levels are associated with a greater acceptance of grey water reuse. Awareness programmes are also important for enhancing the knowledge and the acceptance of the public regarding reclaimed water usage.

According to Jeffrey (2002), the acceptance to reclaim grey water is a function of the end-use for which it is intended. Furthermore, acceptance is lower for end-uses where contact with grey water is higher (e.g. washing machines) than for end-uses with minimal contact (e.g. toilets flushing).From the literature review presented herein, it is clear that the use of reclaimed water increases in regions under water scarcity. Moreover, there must be an expansion of public awareness campaigns in order to promote the benefits of alternative water sources. An expansion of researches is also important so as to spell out issues related to the social acceptance of alternative water sources, as well as water consumption patterns in houses. This, in turn, will help the development of policies on the use of alternative water sources in residential buildings.

The objective of this chapter is to analyse the acceptance of alternative water sources, the water consumption patterns, and the water end-uses in low-class houses in Florianópolis, Southern Brazil.

METHOD

The method applied in this chapter consisted of interviews and assessments of public acceptance on alternative water sources, water consumption patterns, and water end-uses in houses. Questionnaires were used in order to determine the socioeconomic characteristics of households, their acceptance to use alternative water sources (i.e. rainwater and grey water), and their willingness to adopt water conservation strategies in their houses. Correlations among such variables were performed through statistical analysis. Water end-uses and water consumption patterns were also estimated. An assessment of the characteristics of existing plumbing fixtures and appliances in the houses was also carried out. Such variables were treated through confidence intervals.

SOCIOECONOMIC CHARACTERISTICS OF HOUSEHOLDS

The influence of socioeconomic characteristics of households on the acceptance of alternative water sources and water consumption patterns in low-class houses were assessed.

The households selected for the study met at least one of the following criteria:

- Household monthly income less than or equal to 3 minimum wages (3 x R$ 622.00 = US$ 987.30 in April 2012);
- House located in a low-class area (e.g. slums, shanty towns, or suburbs with a high concentration of low-class houses);
- Householder own residence funded by the programme *Minha Casa Minha Vida* or other Brazilian public housing programme for low-income households.

Through socioeconomic questionnaires, the characteristics of households were determined. The data encompassed in the questionnaires included the characteristics of each household member, including: name, gender, date of birth or age, nationality, education, time residing in the house, profession, and monthly income.

Furthermore, general characteristics of households were also determined as outlined in Table 1. The average result for households socioeconomic data are presented in the results section. Such data are also used for analysing the public acceptance of alternative water sources.

Table 1. General characteristics of households

1.Type of house: ( ) rented – R$_______, ( ) free (concession), ( ) owned: ( ) without mortgage; ( ) with mortgage.
2. Water metering system: ( ) communal water meter, ( ) household water meter, ( ) no metering.
3. Water shortages during Summer time? ( ) always, ( ) eventually, ( ) rarely, ( ) never.
4. Water shortages during Winter time? ( ) always, ( ) eventually, ( ) rarely, ( ) never.
5. Water leakages in the buildings? ( ) many, ( ) some, ( ) few, ( ) none. How long has it been leaking? ____ (months)
6. Sewerage system in the building: ( ) on-site treatment (e.g. septic tank), ( ) without treatment, ( ) centralised public treatment system.
7. The town water supply quality is typically: ( ) very good, ( ) good, ( ) regular, ( ) bad, ( ) very bad.
8.Is the water service rate expensive? ( ) No, ( ) Yes.
9.Do you know the water rates for low-class households? ( ) No, ( ) Yes

ANALYSIS OF THE PUBLIC ACCEPTANCE OF ALTERNATIVE WATER SOURCES

The analysis of public acceptance of alternative water sources was carried out through interviews in 44 low-class households. Through the questionnaire, the acceptance of the interviewees regarding the use of rainwater from roof tops for potable water end-uses (i.e. drinking, cooking, and body hygiene end-uses), as well as for non-potable water end-uses (i.e. washing floor, car and clothes, irrigation, and toilet flushing) was assessed. Interviewers clarified for respondents that the water quality of the harvested rainwater would be improved by diverting the first flush of rainfall events, which carries most of the rainwater pollutants.

The acceptance to use treated grey water for non-potable residential water end-uses was assessed for both building and communal scale reclamation systems. Interviewers explained to respondents that: (i) the raw grey water streams would be treated in constructed wetlands (i.e. sand filter with plants) followed by disinfection through chlorination; (ii) the treated grey water would be used only for non-potable purposes (i.e. use of treated wastewater stream from baths, washing basins, and washing machines); (iii) building scale reclamation systems supply treated grey water only from the household; and (iv) communal scale reclamation systems supply treated grey water from other households as well.

Table 2 shows the questions used to assess the public acceptance to use alternative water sources.

Table 2. Public acceptance of alternative water sources

1. Rainwater would be used for: ( ) drinking, ( ) dish washing, ( ) hands washing, and ( ) showering.
2. Rainwater would be used for: ( ) car washing, ( ) external paved areas washing, ( ) internal floor washing, ( ) clothes washing, ( ) irrigation, and ( ) toilet flushing.
3 Treated grey water from the household would be used for: ( ) external paved areas washing, ( ) internal floor washing, ( ) irrigation, and ( ) toilet flushing.
4. Treated grey water from other households would be used for: ( ) external paved areas washing, ( ) internal floor washing, ( ) irrigation, and ( ) toilet flushing.

Data on the acceptance to use rainwater and grey water were correlated with the number of occupants in each household, total and per capita income, age and level of education. The correlation analysis was performed by using the Pearson method in the software Minitab 16 (Minitab, 2010). As the household sample size was small (n=44) and the variability of the data was large, the confidence interval of 90% was chosen to perform the correlation analysis.

Some variables were considered as continuous, including: household income, per capita income, level of education, age, and number of inhabitants. The other 22 variables were considered as discrete by attributing values equal to zero or one to negative and positive answers, respectively. Table 3 shows all variables used for the correlation analysis.

Table 3. Variables used for the Pearson's correlation analysis

Variable	Description	Type
Household income (R$)	Total household income	Continuous
Per capita income (R$)	Total household income divided by the number of residents	Continuous
Number of residents	Number of residents in the household	Continuous
Age	Average age of residents	Continuous
Education level*	Average level of education achieved by residents	Continuous
Acceptance of rainwater for potable end-uses	Acceptance to use for a range of water end-uses	Discrete
Acceptance of rainwater for non-potable end-uses		Discrete
Acceptance of treated grey water from the household		Discrete
Acceptance of treated grey water from other households		Discrete

*The education level of residents was classified in classes varying from zero (illiterate) to nine (complete tertiary). The education level for each household was equal to the average education level of its residents.

ESTIMATION OF WATER END-USES

In order to determine the fraction of total water demand that could be met by each alternative water source studied, the water end-uses in 48 low-class households were estimated.

Water end-uses were estimated by interviewing householders about their water consumption pattern (i.e. usage time and flow rate) for each plumbing fixture and appliance in the house. Water end-uses were classified as potable and non-potable uses. Potable uses refer to uses that need to be within the standards of potability (i.e. kitchen tap, washing basin and shower). Non-potable water end-uses refer to toilet flushing, laundry and external taps.

The flow rate of taps and showers was determined by asking householders to use such plumbing fixtures as usual for three times. In order to determine the flow rate, the water volume and the duration of each event was registered with the use of containers and timers, respectively. The flow rate of toilets fitted with flushing valves was considered equal to the values suggested by the Brazilian plumbing code NBR 5626 (ABNT, 1998). The average volume per use cycle of toilets fitted with water cisterns and washing machines was assumed to be equal to values presented by manufacturers and in the PROCEL (Brazilian Energy Efficiency Programme) catalogue. The average flow rate of each tap and shower was estimated using Equation 1. The equipment characterized by time of use are: shower, washing basin, kitchen tap, external tap and laundry tap. The devices characterized by cycle are toilet with flushing valve or toilet with cistern and washing machine.

$$Q_{ij} = \sum_{k=0}^{3} \frac{q_{ijk}}{3} = \sum_{k=0}^{3} \frac{V_{ijk}}{3 \times t_{ijk}} \quad (1)$$

where: Q_{ij} is the flow rate of each plumbing fixture *i* (L/s) for each household *j*; q_{ijk} is the flow rate (L/s) for each event out of the three measurements *k*, for each plumbing fixture *i* in each household *j*; V_{ijk} is the volume (L) in each measurement *k*, for each plumbing fixture *i* in each household *j*; $t_{ijk}$is the time (s) registered in each event *k*, for each plumbing fixture *i* in each household *j*.

The operation pattern of each plumbing fixture or appliance was estimated through their frequency of use. By asking the householders, the frequency of use was registered for each hour of the day for each plumbing fixture or appliance. For plumbing fixtures which had the flow rate estimated, the duration of use in each hour was estimated; while, for appliances which had the volume per cycle estimated, the number of cycles in each hour was estimated. Moreover, the weekly and monthly frequency of use were also estimated.

Depending on the plumbing fixture or appliance characteristic, the hourly water consumption was estimated either by using the hourly usage time and the average flow rate, or by using the hourly number of cycles and volume per cycle. The monthly water consumption for each plumbing fixture and appliance was estimated by multiplying their frequency of use (i.e. daily, weekly, or monthly) by the estimated daily water consumption. The water end-uses were calculated by using Equation 2.

$$UF_i = \frac{100 \times V_i}{\sum_{i=0}^{n} V_i} \quad (2)$$

where: UF_i is the water end-use (%) of each plumbing fixture or appliance *i* in each household; V_i is the water consumption (m^3/month) of the plumbing fixture or appliance *i* in the household; *n* is the number of plumbing fixtures or appliance in the household.

Water end-uses were presented as the mean value and the 90% confidence interval for two groups among the studied households. The two groups were separated by household income, in which the first had income less or equal to three minimum wages, and the second within three and five minimum wages. The households in the second group were either located in low-class suburbs, or were funded by low-class housing programmes.

So as to determine the confidence interval, the Student's t-test distribution considering a degree of freedom equal to one value less than the sample size (i.e. n-1) was used. The calculations were performed using the software Minitab 16 (Minitab, 2010).

Characteristics of Plumbing Fixtures and Appliances and Water Consumption Patterns

The average and 90% confidence interval water flow rate or volume per cycle of each plumbing fixture or appliance was determined for the 48 households. The average operation pattern of each plumbing fixture or appliance in time or number of cycles per hour was estimated for the 11 households. The confidence interval for the flow rate was also estimated by means of the Student's t-test distribution.

The average hourly water consumption pattern for showers, taps and toilets with flushing valves was estimated by using the calculated average flow rate and the average usage time; whereas, for washing machines and toilets with cisterns, it was used the calculated average volume per cycle and the average hourly number of cycles. The water consumption pattern for each fixture represents a day with mean monthly water consumption.

RESULTS

SOCIOECONOMIC CHARACTERISTICS

Figure 3 illustrates the socioeconomic characteristics of the studied households. The median household and per capita incomes were approximately equal to R$ 2000.00 and R$ 650.00 (US$ 993.20 and US$ 322.79), respectively; whereas the maximum household and per capita incomes were equal to R$ 8000.00 and R$ 4000.00 (US$ 3972.79 and US$ 1986.39), respectively.

These maximum values are likely outliers as the average values are considerably lower. The number of occupants ranged from 1 to 6, with an average value of 4 occupants. The average age was equal to 39 years old. The average education level was equal to 4, which is equivalent to incomplete primary education.

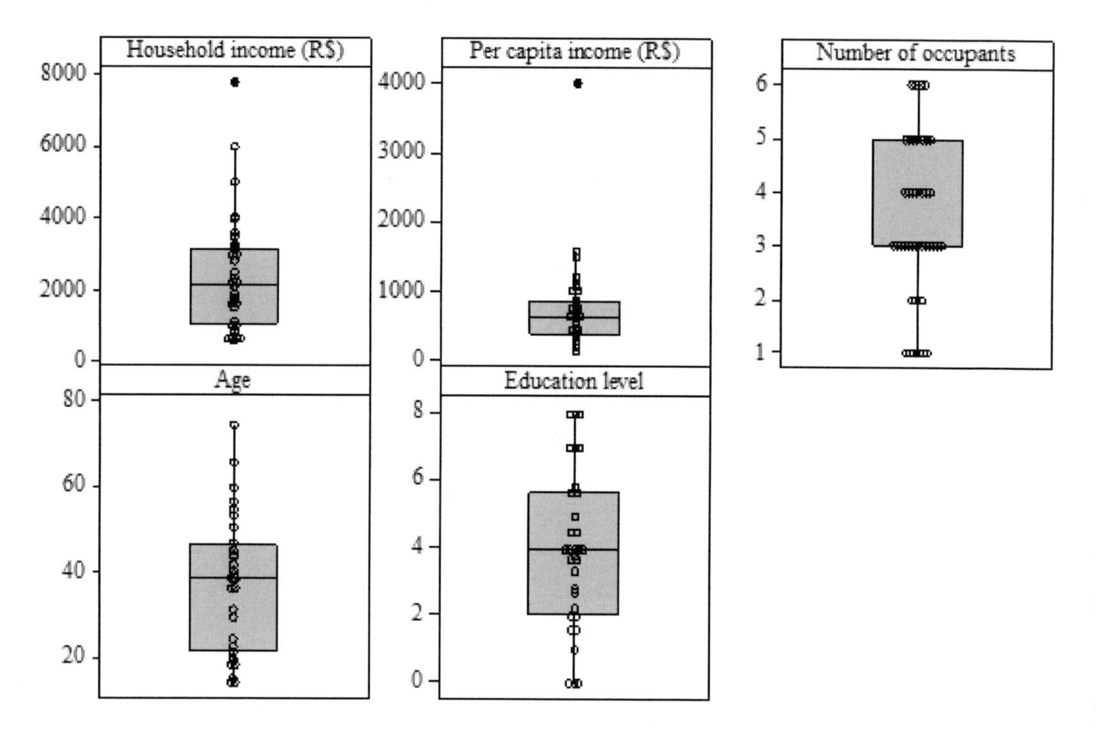

Figure 3. Socioeconomic characteristics of studied households.

ACCEPTANCE OF ALTERNATIVE WATER SOURCES

Figure 4 shows the public acceptance to use alternative water sources for different water end-uses in the studied households. The percentage of acceptance to use treated rainwater for potable water end-uses ranged between 27 and 59%. Rainwater had even greater acceptance for non-potable water end-uses, ranging from 84 to 95% for washing machine and toilet flushing, respectively.

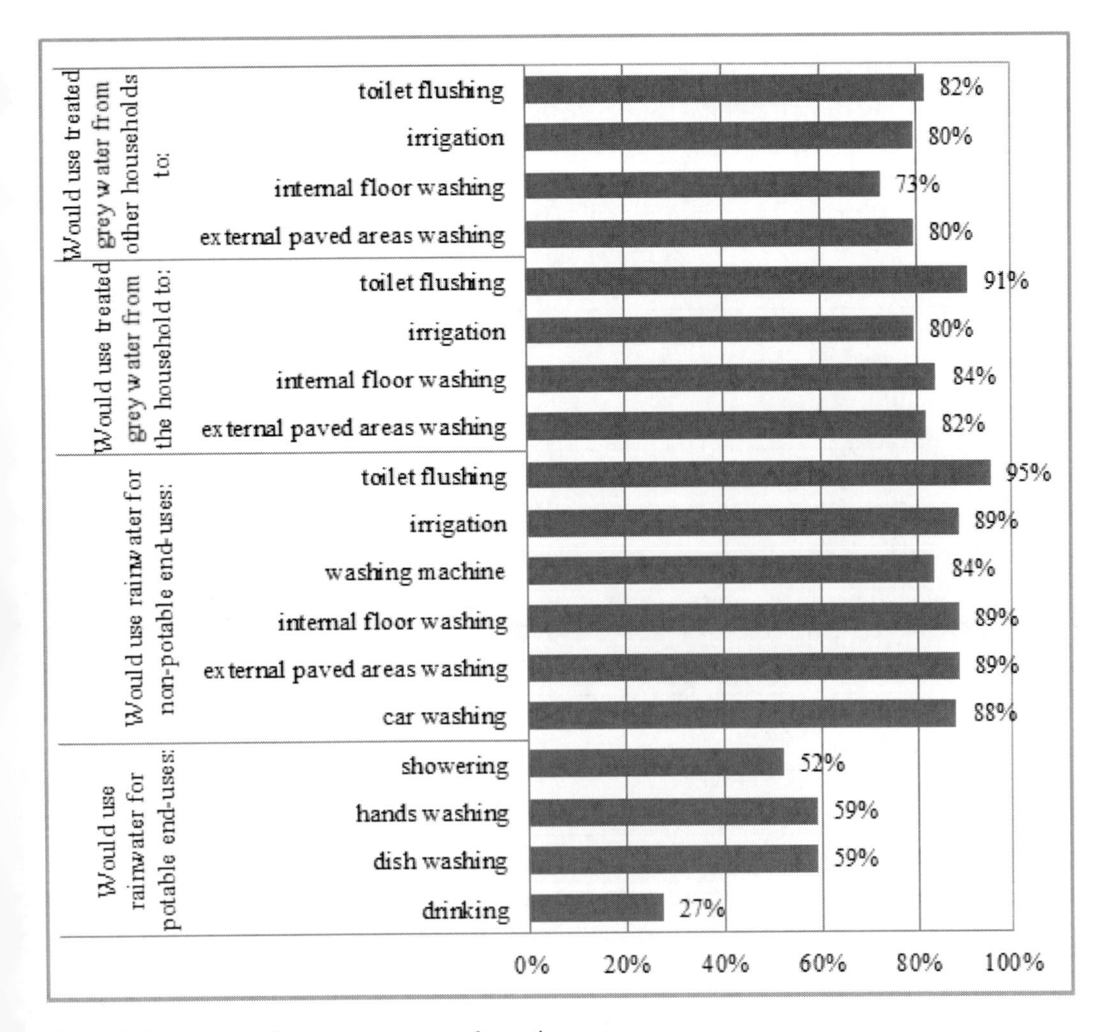

Figure 4. Percentage of acceptance to use alternative water sources.

The acceptance to use treated grey water from within the household for non-potable purposes varied from 80 to 91%. For treated grey water generated in other households, the acceptance slightly decreased to a range between 73 and 82%. The acceptance varied in accordance to the contact with grey water, in which toilet flushing and irrigation were uses more acceptable. It was observed that there is a greater occupant acceptance when the grey water comes from his/her own house, as opposed to the alternative coming from neighboring households.

Table 4 presents the results for the perception of householders about local centralised water and wastewater services. Most of the studied houses were owned by the householders. Their water consumption was predominantly measured by individual meters. As for the water supply reliability, 25% of respondents declared that they always undergo water shortage during summer. During winter, 9% of respondents always undergo water shortage, while 44% never undergo water shortages over this season.

The wastewater was treated using septic tanks and centralised wastewater services for 57% and 40% of the studied households, respectively. The water quality of the town water supply was considered good by 58% of the respondents. In 80% of the households, no water leakage problems were reported. The water rates charged by the local water utility were considered expensive by 41% of the householders; however, only 23% of them were aware of the existence of low-class water charge rates.

Table 4. Householders occupation, water metering, and perception of local water and wastewater services

Household occupation					
Option	rented	free	owned		
Answer	16%	5%	79%		
Water metering system					
Option	community	individual	none		
Answer	40%	58%	2%		
Is there lack of water in the summer?					
Option	always	sometimes	seldom	never	
Answer	25%	30%	20%	25%	
Is there lack of water in the winter?					
Option	always	sometimes	seldom	never	
Answer	9%	19%	28%	44%	
Are there water leakages?					
Option	many	some	few	none	
Answer	9%	4%	7%	80%	
Treatment of sewage: (*septic tank)					
Option	local*	without treatment	public network		
Answer	57%	2%	40%		
Quality of water					
Option	very good	good	regular	bad	too bad
Answer	5%	58%	30%	8%	0%
Is the water rate expensive?					
Option	yes	no			
Answer	41%	59%			
Know the social rate for water?					
Option	yes	no			
Answer	23%	77%			

Tables 5 and 6 show the results for the Pearsons's correlation analysis at 90% confidence interval. These tables show only results for variables with significant correlation between socioeconomic parameters and the acceptance of studied alternative water sources for different water end-uses; correlations without significance were not presented or replaced by "--".

Table 5. Correlation between acceptance of alternative water sources for different end-uses and age and education level

Variable	Data	Age (years)	Education level
Education level	Pearson's	0.459	--
	p-value	0.006	--
Acceptance to use treated rainwater for drinking	Pearson's	--	-0.322
	p-value	--	0.059
Acceptance to use treated grey water from the household for irrigation	Pearson's	-0.365	-0.321
	p-value	0.031	0.060
Acceptance to use treated grey water from the household for at least one water end-use	Pearson's	-0.349	--
	p-value	0.040	--

Table 6. Correlation between acceptance of alternative water sources for different end-uses and number of residents and income

Variable	Data	Household income (R$)	Per capita income (R$)	Number of residents
Number of residents	Pearson's	0.439	--	--
	p-value	0.004	--	--
Per capita income (R$)	Pearson's	0.557	--	-0.323
	p-value	0.000	--	0.037
Age (years)	Pearson's	0.292	0.390	--
	p-value	0.094	0.023	--
Acceptance to use rainwater for hand washing	Pearson's	--	0.335	--
	p-value	--	0.030	--
Acceptance to use rainwater for car washing	Pearson's	0.283	--	--
	p-value	0.070	--	--
Acceptance to use grey water from other households for irrigation	Pearson's	0.272	--	0.285
	p-value	0.081	--	0.061
Acceptance to use grey water from other households for toilet flushing	Pearson's	0.261	--	0.282
	p-value	0.095	--	0.064

The Pearson's correlation coefficient indicates the significance of the correlation between studied variables. The larger the absolute value of the coefficient, the higher the correlation, where positive and negative values indicate a direct and an inverse correlation, respectively. Therefore, for positive correlations, an increase of the socioeconomic variable leads to an increase in the public acceptance.

Rainwater use for drinking purposes was correlated to the level of education, with 90% confidence. The lower the education level of householders, the greater the acceptance was. This is probably due to the general characteristics of rainwater (i.e. low colour and turbidity), which possibly makes the public think it is clean, and hence suitable for drinking. However, by attending school, people learn about the importance of water treatment for reducing water born diseases, which is typically achieved by centralised water treatment systems. Likely, the more educated group have the perception that potable water standards can only be achieved by means of centralised water systems.

Similarly to rainwater acceptance, grey water acceptance for irrigation purposes was inversely correlated with the education level, with 90% confidence. It is believed that the formal education has influenced again in the perception of people. Therefore, people with formal education probably were concerned with health issues, despite the intuitive perception expressed by the people with low education that their own treated grey water may bring minor risks for their health. Probably, the formal education limits people to think about the feasibility of decentralised systems, as centralised systems are the typical approach undertaken in developed regions.

The age of householders was also inversely correlated with the acceptance to use grey water for non-potable end-uses, with 90% confidence. Possibly, this is due to a shift in the way people perceive the development of houses and cities in the recent past. It is likely that young generations are more prone to learn and adhere to new sustainable development practices (e.g. the use of alternative water sources), as they have typically more access to new information and are not based on past knowledge.

The use of rainwater for car washing was directly correlated with the household income. Certainly, wealthier householders have cars, while poorer householders do not. Therefore, the first group could accept more rainwater for car washing as they are more likely to use water for this purpose, while, for the second group, it may not be the case. Possibly, similar aspects may have also influenced the positive correlation between hand washing and per capita income.

The acceptance of treated grey water from other households for irrigation and toilet flushing was correlated to the total household income. This may be influenced by the higher acceptance that wealthier householders had for rainwater. Possibly, the introduction of one alternative water source may influence the acceptance to other alternative water sources.

Grey water for irrigation and toilet flushing end-uses was also correlated with the number of householders. Possibly, smaller households are more concerned with health risks from the contact with different people than larger households, who may be visited by a larger number of foreign people, and, hence, be less concerned with the contact with them.

The acceptance to use rainwater for dish washing and showering was not correlated to socioeconomic variables. It can be noticed that the education level and population age influence the acceptance to the reuse of grey water. In general, the higher the user acceptance, the lower the education level and age. As for the income, the higher the income, the higher the acceptance.

WATER END-USES

The water end-uses of 48 households in the metropolitan region of Florianópolis, Southern Brazil, were estimated by using the flow rates and volume per cycle shown in Tables 7 and 8, respectively. The flow rate of plumbing fixtures varied considerably among households as observed by their large standard deviation (Table 7). Such variation was greater for end-uses that are generally operated at their maximum flow rates, i.e. laundry and external taps, which presented average flow rates of 0.13 and 0.17 L/s, respectively. These fixtures are typically operated at their maximum capacity as the greater the flow rate is, the quicker the intended function will be accomplished (e.g. filling up a bucket or irrigating plants). Therefore, the larger variation in the flow rate of such equipment may be directly proportional to the maximum flow rate that can be delivered depending on the characteristics of the plumbing systems, including but not limited to: (i) direct or indirect town water supply; (ii) head losses in the plumbing components; (iii) tap type.

A similar trend was observed for the volume per cycle of plumbing fixtures and appliances (Table 8). Such parameter presented a larger variation for washing machines as their volume is directly proportional to the washing machine type and model and washing cycle used.

Table 7. Average and standard deviation of the flow rate for plumbing fixtures

Plumbing fixture	Average flow rate (L/s)	Standard deviation (L/s)
Shower	0.07	0.03
Washing basin	0.08	0.04
Kitchen tap	0.09	0.05
Laundry tap	0.13	0.13
External tap	0.17	0.11

Table 8. Average and standard deviation of the volume per cycle for plumbing fixtures and appliances

Plumbing fixture or appliance	Average water consumption (L/cycle)	Standard deviation (L/cyle)
Toilet with flushing valve	6.2	1.4
Toilet with cystern	11.5	2.6
Washing machine	102.5	50.2

The water consumption patterns per capita between the two studied income groups were not significantly different. Households with income up to three minimum wages consumed on average 153 L/capita/day and 374 L/household/day; whereas households with income between three and five minimum wages consumed on average 164 L/capita/day and 616 L/household/day. Similar water consumptions (per capita) were found in other studies (Oliveira et al., 2006; Dantas et al., 2006; Ywashima et al., 2006)

The total monthly water consumption was lower than 10 m³ for 33% of the 48 households studied. The water consumption pattern was associated with the household occupation pattern, in which the longer the permanence of residents in the house, the greater the water consumption was. Similar results were described by Vieira (2012) for low income households in Florianópolis. Table 9 shows the monthly water consumption for each plumbing fixture and appliance, in which the average and the upper and lower limit values are presented.

The washing machine had an average water consumption, lower and upper limit of 0.49, 0.39 and 0.59, respectively. Table 10 shows the water end-uses for each plumbing fixture and appliance.

Table 9. Average, lower and upper limit values for the monthly water consumption in plumbing fixtures and appliances

Plumbing fixture or appliance	Water consumption (m^3/capita/month)	Standard deviation (m^3/capita/month)	Confidence interval 90%	
			Lower limit	Upper limit
Shower	1.49	1.17	1.28	1.71
Toilet flushing	0.89	0.57	0.78	0.99
Kitchen tap	0.82	0.54	0.72	0.92
Washing machine	0.49	0.48	0.39	0.59
External tap	0.31	0.26	0.21	0.40
Laundry tap	0.28	0.43	0.14	0.41
Washing basin	0.24	0.26	0.18	0.30
Others	0.06	0.08	0.04	0.07

Table 10. Water end-uses

Plumbing fixture or appliance	Water end-uses (%)	Confidence interval 90%	
		Lower limit (%)	Upper limit (%)
Shower	32.7	27.9	37.5
Toilet flushing	19.4	17.0	21.8
Kitchen tap	18.0	15.8	20.2
Washing machine	10.7	8.5	13.0
External tap	6.7	4.6	8.8
Laundry tap	6.1	3.1	9.1
Washing basin	5.2	3.9	6.6
Others	1.2	0.9	1.6

The largest water end-use (32.7%) was for showering. This end-use was followed by toilet flushing (19.4%) and dish washing (18.0%). These three end-uses accounted for 70.1% of the total water consumption in households on average. Nonetheless, the water consumption pattern for each plumbing fixture and appliance varied significantly among sampled households due to resident's different consumption habits.

By using confidence intervals, the variability of the water consumption pattern for each appliance was estimated. To exemplify this, showering had an average consumption of 1.49 m^3/capita/month, but a lower and an upper limit of 1.28 and 1.71 m^3/capita.month, respectively. The average water end-use for this fixture was 32.7%, and the lower and upper limits equal to 27.9 and 37.5%, respectively.

Kitchen tap and toilet flushing had a similar monthly water consumption pattern. Both end-uses were not significantly different as per the T-test for 90% confidence interval, as they presented a large variation among studied households.

Non-potable water end-uses (i.e. toilet flushing, washing machine, and laundry and external taps) were equivalent to 42.9% of the total water consumption in households on average, ranging from 10.3 to 60.0%. Such end-uses can be supplied with alternative water sources, including rainwater for all end-uses thereof. Grey water for toilet flushing and external taps could be used to supply 26.1% of the total water demand in households; whereas it could be produced from 54.7% of the wastewater streams generated in the households from showers, washing basins, washing machines, and laundry tap. Therefore, this water source has a great potential for use as the supply capacity of grey water surpasses the demand.

WATER CONSUMPTION PATTERNS

The daily average water consumption patterns estimated for each plumbing fixture and appliances in the 11 households studied are presented in Figures 5-10. Figure 5 shows that the showers present two consumption peaks during the day, one at 8h and another at 20h. On the other hand, toilets water consumption peaked only once per day at 22h (Figure 6). Kitchen taps had the largest hourly consumption at 11h and 13h. Such usage pattern in the kitchen may suggest that householders have lunch at home, what may increase the household water consumption. Similarly to the analysis of the water end-uses, analysis of the average daily water consumption patterns also showed that the largest consumption of water occurs in the shower, toilet flushing and kitchen tap.

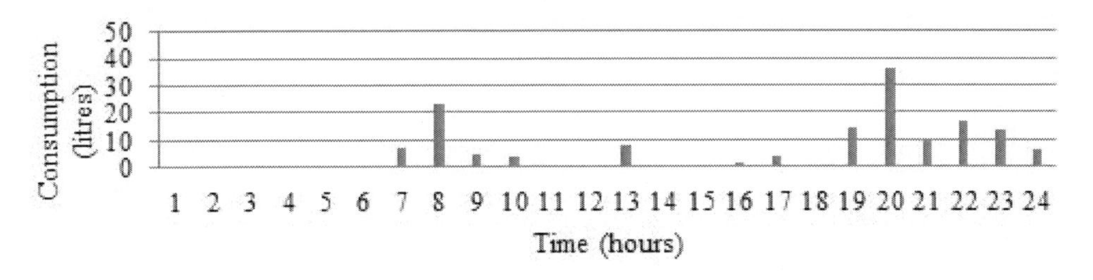

Figure 5 – Average hourly water consumption pattern for showers.

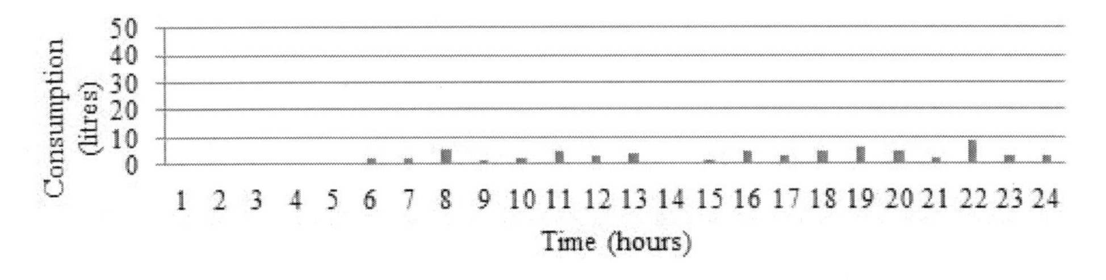

Figure 6. Average hourly water consumption pattern for toilets.

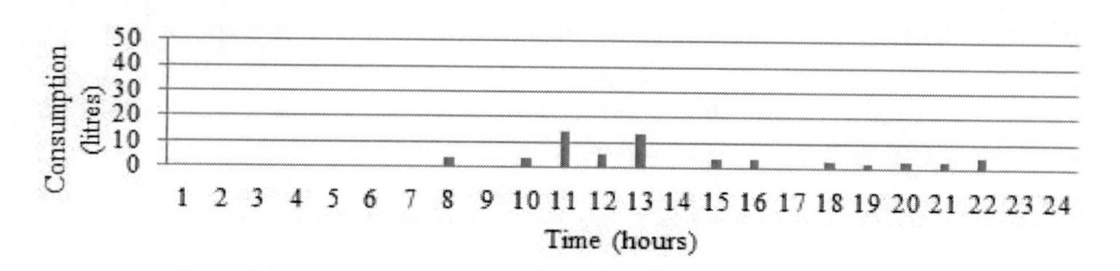

Figure 7. Average hourly water consumption pattern for kitchen taps.

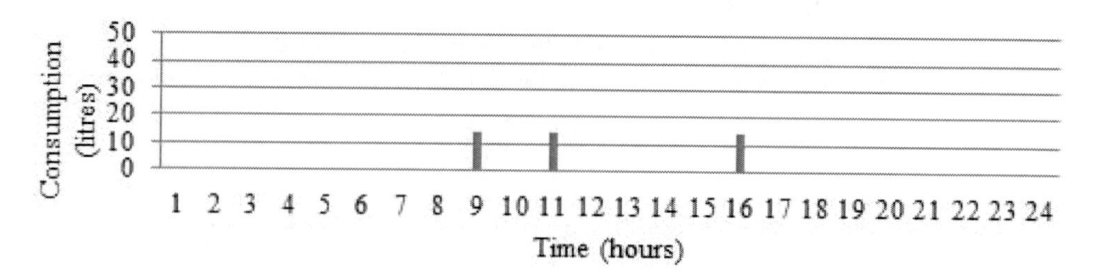

Figure 8. Average hourly water consumption pattern for washing machines.

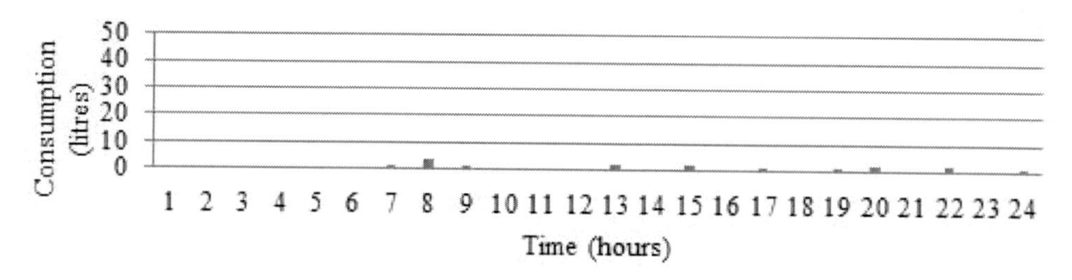

Figure 9. Average hourly water consumption pattern for washing basins.

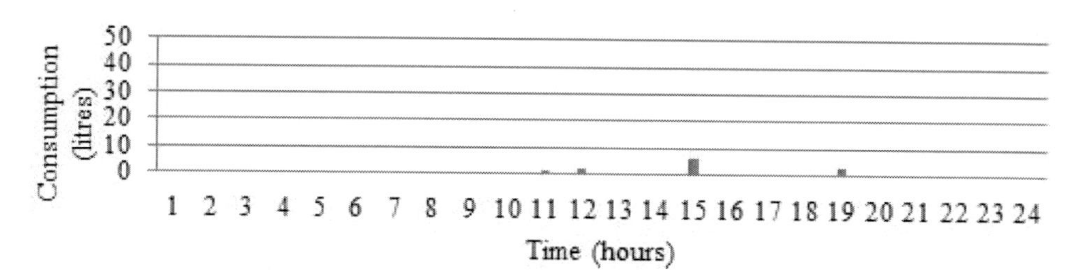

Figure 10. Average hourly water consumption pattern for laundry taps.

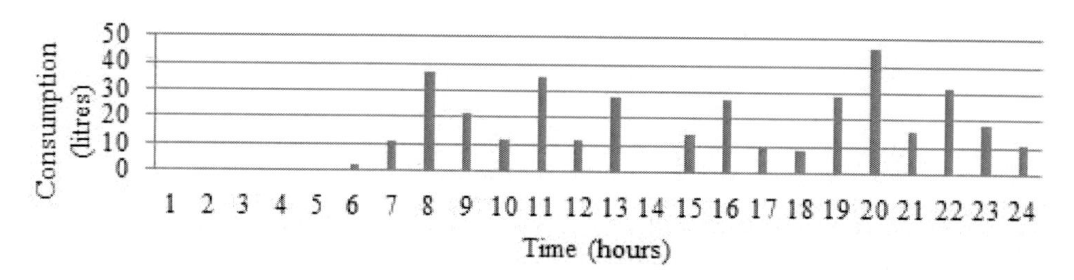

Figure 11. Average hourly water consumption pattern for the whole house plumbing fixtures and appliances.

CONCLUSION

The use of alternative water sources depends on several factors, such as economic feasibility, technical, environmental, water quality and social acceptance. Among these factors, the social acceptance is critical to the successful implementation of alternative sources in a community. This chapter addressed this issue by performing interviews in low-class houses in Southern Brazil. The acceptance of alternative water sources was attained by interviewing 44 householders.

The acceptance results showed that from 27 to 59% of the householders would use rainwater for potable end-uses, and from 84 to 96% would use it for non-potable end-uses. Toilet flushing had the greatest acceptance for rainwater usage. From 80 to 91% of the householders would use grey water from their own houses for non-potable end-uses, and from 73 to 82% would use it originated from other houses. In an attempt to understand these results, the correlation statistics showed that the older people are, the less they accept grey water as an alternative water source; the greater the degree of instruction, the less the acceptance of rainwater; the greater the number of occupants in the house, the less the acceptance of rainwater for washing dishes.

The water end-uses were assessed in 48 low-class households in order to determine the percentage of the total water demand that could be met by alternative water sources. Non-potable water end-uses were equal to 42.9% of the total water demand at households on average. Results show that the shower has the greatest water end-use with 32.7% of representativeness, followed by the toilet flushing with 19.4% and the kitchen tap with 18.0%.

The pattern of daily water consumption was estimated in 11 households. Showers, kitchen taps and toilet flushing showed the higher daily consumption of water as well as water end-uses.

The results of this chapter may be used to improve the forecast of future water demand and to implement water efficiency programmes in low-class houses. Such data may be also used in awareness campaigns about water conservation practices and alternative water sources use.

Likely, there would be a deferral of capital investments in new water assets for enhancing water and wastewater services by saving water in low-class houses. In order to achieve a reduction of water consumption in low-class houses, awareness campaigns are important for improving both the knowledge of householders about new technologies, and their acceptance to use alternative water sources.

In this context, knowledge about the public acceptance of alternative water sources in low-class households, water end-uses, and water consumption patterns are crucial for fostering water conservation practices.

REFERENCES

ABNT. Associação Brasileira de Normas Técnicas. (1998). NBR 5626: Instalação predial de água fria [Brazilian plumbing code]. Rio de Janeiro. [in Portuguese]

Almeida, G. (2007). Metodologia para caracterização de efluentes domésticos para fins de reuso: estudo em Feira de Santana, Bahia [Method to characterize domestic wastewater streams for reclamation: study in Feira de Santana, Bahia]. Dissertação de Mestrado [Masters thesis] (Gerenciamento e Tecnologia Ambiental no Processo Produtivo). Escola Politécnica, Universidade Federal da Bahia. [in Portuguese]

Boyer, T. H., Overdevest, C., Christiansen, L., Ishii, S. K. L. (2012). Expert stakeholder attitudes and support for alternative water sources in a groundwater depleted region. *Science of the Total Environment*, 437, 245–254.

Cohim, E., Cohim, F. (2007). Reuso de água cinza: a percepção do usuário (estudo exploratório) [Grey water reclamation: a perception of users (exploratory study)]. II - 418. 11 p. In: Anais do 24º Congresso Brasileiro de Engenharia Sanitária e Ambiental. Belo Horizonte. CD-ROM. [in Portuguese]

Dantas, C. T.; Ubaldo Jr., L.; Potier, A. C.; Ilha, M. S. O. (2006). Caracterização do uso de água em residências de interesse social em Itajubá [Water consumption assessment in low-class households]. XI Encontro Nacional de Tecnologia do Ambiente Construído, Florianópolis – SC, Anais... CD Rom. [in Portuguese]

Dias, I. C. S., Athayde Júnior, G. B., Gadelha, C. L. M. (2007). Viabilidade econômica e social do aproveitamento de águas pluviais em residências na cidade de João Pessoa. X Simpósio Nacional de Sistemas Prediais: Desenvolvimento e inovação [Economic and social feasibility for rainwater use at households in João Pessoa]. São Carlos. [in Portuguese]

Dolnicar, S., Hurlimann, A., Grün, B. (2011). What affects public acceptance of recycled and desalinated water? *Water Research,* 45, 933–943.

Dolnicar, S., Hurlimann, A., Nghiem, L. D. (2010). The effect of information on public acceptance e The case of water from alternative sources. *Journal of Environmental Management,* 91, 1288-1293.

Domènech, L., Saurí, D. (2010). Socio-technical transitions in water scarcity contexts: Public acceptance of greywater reuse technologies in the Metropolitan Area of Barcelona. *Resources, Conservation and Recycling*, 55, 53–62.

Friedler, E., Lahav, O., Jizhaki, H., & Lahav, T. (2006). Study of urban population attitudes towards various wastewater reuse options: Israel as a case study. *Journal of Environmental Management*, *81*, 360-370.

Garcia, A. P. A. (2011). Fatores associados ao consumo de água em residências de baixa renda. Dissertação de Mestrado (Engenharia Industrial) [Factors associated with water consumption in low-class households]. Escola Politécnica, Universidade Federal da Bahia. [in Portuguese]

Gonçalves; R. F., Silva, L. M. (2012). Aspectos legais e normativos relativos ao reuso / reaproveitamento de água para fins não potáveis [Legal requirements for non-potable water reclamation]. 3º Work Shop - Rede de pesquisa - Uso racional da água e eficiência energética em habitação de interesse social. [in Portuguese]

Hafner, A. V. (2007). Conservação e reuso de água em edificações - experiências nacionais e internacionais [Water conservation and reclamation in buildings – national and international experiences]. Dissertação de Mestrado (Engenharia Civil). Universidade Federal do Rio de Janeiro. [in Portuguese]

Hurlimann, A., Dolnicar, S. (2012). Newspaper coverage of water issues in Australia. *Water Research*, 46, 6497–6507.

Hurlimann, A., Dolnicar, S. (2010). When public opposition defeats alternative water projects – The case of Toowoomba Australia. *Water Research,* 44, 287–297.

Hurlimann, A., Dolnicar, S., Meyer, P. (2009). Understanding behaviour to inform water supply management in developed nations – A review of literature, conceptual model and research agenda. *Journal of Environmental Management*, 91, 47–56.

Hurlimann, A., & Mckay, J. (2007). Urban Australians using recycled water for domestic non-potable use — An evaluation of the attributes price, saltiness, colour and odour using conjoint analysis. *Journal of Environmental Management, 83*, 93-104.

Jamrah, A.; Al-Futaisi, A.; Prathapar, S. A.; Ahmad, M.; Al Harasi, A. (2004). *International Conference Water Demand Management*, Dead Sea, Jordan.

Jeffrey, P. (2002). Public attitudes to in-house water recycling in England and Wales. *Journal of the Chartered Institution of Water and Environmental Management*, 16, 214-217.

Mancuso, P. C. S. (2003). "Tecnologia de reúso de água" ["Technology for water reclamation"]. In: Mancuso, P.C.S., Santos, H.F, Reúso de Água, 1.ed. cap. 9, Barueri, SP, Editora Manole. [in Portuguese]

Mankad, A., Greenhill, M., Tucker, D., Tapsuwan, S. (2013). Motivational indicators of protective behaviour in response to urban water shortage threat. *Journal of Hydrology*. Available online. In Press.

Mankad, A. (2012). Decentralised water systems: Emotional influences on resource decision making. *Environment International*, 44(1), 128-140.

Mankad, A., Tapsuwan, S. (2011). Review of socioeconomic drivers of community acceptance and adoption of decentralised water systems. *Journal of Environmental Management*, 92, 380-391.

Martinetti, T. H., Shimbo I., Teixeira, B. A. N. (2007). Análise de alternativas mais sustentáveis para tratamento local de efluentes sanitários residenciais [Environmental feasibility analysis for domestic wastewater treament on site]. X Simpósio Nacional de Sistemas Prediais: Desenvolvimento e inovação. São Carlos. [in Portuguese]

Minitab (2010). Minitab 16 Statistical Software. [Computer software]. State College, PA: Minitab, Inc. (<www.minitab.com>)

Morales-Pinzón, T., Lurueña, R., Rieradevall, J.,Gasol, C.M, Gabarrel, X. (2012). Financial feasibility and environmental analysis of potential rainwater harvesting systems: A case study in Spain. *Resources, Conservation and Recycling*, 69, 130-140.

Mourad, K. A., Berndtsson, J. C., Berndtsson, R. (2011). Potential fresh water saving using greywater in toilet flushing in Syria. *Journal of Environmental Management,* 92, 2447-2453.

Muthukumaran, S., Baskaran, K., & Sexton, N. (2011). Quantification of potable water savings by residential water conservation and reuse – A case study. *Resources, Conservation and Recycling*, *55*(11), 945-952.

Oliveira, L. H.; Sousa, L. C.; Silva, K. A.; Paixão, A. (2006). Caracterização do uso da água em habitações unifamiliares de interesse social [Water consumption assessment in single-family low class households]. XI Encontro Nacional de Tecnologia do Ambiente Construído, Florianópolis-SC, Anais... CD Rom. [in Portuguese]

Po, M., Kaercher, J. D., Nancarrow, B. E. (2003). Literature review of factors influencing public perceptions of water reuse. CSIRO Land and Water. Technical report 54/03, December.

Ryan, A. M., Spash, C. L., & Measham, T. G. (2009). Socioeconomic and psychological predictors of domestic greywater and rainwater collection: Evidence from Australia. *Journal of Hydrology*, *379*(1-2), 164-171.

Santos, D. C. dos, Froehner, S. (2007). Qualidade da água pluvial e da água cinza nas edificações: estudo comparativo [Rainwater and grey water quality in buildings: a comparative study]. X Simpósio Nacional de Sistemas Prediais: Desenvolvimento e inovação. São Carlos. [in Portuguese]

Vieira, A. S. (2012) Uso racional de água em habitações de interesse social como estratégia para a conservação de energia em Florianópolis, Santa Catarina [Integrated water management in low-income households for energy conservation in Florianópolis city]. Dissertação de Mestrado (Engenharia Civil) [Masters thesis]. Universidade Federal de Santa Catarina. [in Portuguese]

Ywashima, L. A.; Campos, M. A. S.; Piaia, E.; Luca, D. M. P.; Ilha, M. S. O. (2006). Caracterização do uso de água em residências de interesse social em Paulínia [Water consumption assessment in low-class households in Paulínia]. XI Encontro Nacional de Tecnologia do Ambiente Construído, Florianópolis-SC, Anais... CD Rom. [in Portuguese]

In: Water Conservation
Editor: Monzur A. Imteaz
ISBN: 978-1-62808-993-6

Chapter 6

CHALLENGES AND POTENTIALS OF REUSING GREYWATER TO REDUCE POTABLE WATER DEMAND

Cristina Santos[1*] and Cristina Matos[2]
[1]Faculdade de Engenharia, Universidade do Porto. Portugal
[2]Science and Technology School, University of Trás-os-Montes e Alto Douro. Portugal

ABSTRACT

Greywater reuse is a very efficient way to reduce potable water consumption in buildings once it promotes a second use of water in appliances that don't need potable water quality, such as toilet flushing, irrigation and pavement washing. It also brings a significant reduction of wastewater produced in urban areas, minimizing the negative impacts of treated wastewater discharges and the water and wastewater treatment costs. This chapter intends to describe greywater reuse systems, presenting the definition of greywater and its main characteristics. Then, a description of greywater reuse systems is presented, describing all types of systems, from the most basic to the most complex ones and the integration of rainwater harvesting in greywater reuse systems Many examples of systems that are implemented in several countries is also presented, including information about installation costs and water savings. This chapter also includes the results of many surveys made in different countries, describing public acceptance of greywater reuse, main concerns and public attitude in places where reuse schemes have already been put in place. Finally, in the last section presents some conclusions about the covered subjects, including the main challenges of this practice, and how to overcome some obstacles to have greywater reuse systems implemented at a large scale in the future.

Keywords: Greywater reuse; sustainable water consumption; public acceptance

---

[*] Faculdade de Engenharia da Universidade do Porto, Rua Dr. Roberto Frias, s/n, 4200-465 Porto, Portugal, Email: csantos@fe.up.pt.

INTRODUCTION

The importance of water for human life and social development is unquestionable and the main activities that depend on it are also the ones that contribute mostly to its degradation. In Europe, an uneven distribution of precipitation and runoff, spatially and temporarily specially in Mediterranean countries, requires the construction of costly water storages and higher levels of wastewater treatments (Marecos do Monte, 1996). Furthermore, seasonal variability in the occupation of the territory leads to a significant stress in coastal areas and requires the deviation of significant volumes of water. So, main problems in some of these countries are the high cost of providing water available at the right place, at the right time with the right quality and not its scarcity (Marecos do Monte, 1996; PNUEA, 2001). Water stress is already a reality in many of these countries and climate change will only accentuate the frequency and intensity of those events in the future, especially in southern European countries (EEA, 2009).

Over decades, water resources are being intensively over exploited and polluted, and it is estimated that, in a few years, high magnitudes of water stress will be observed not only in Europe but also in many countries around the world. To reverse the non-sustainable tendency of increasing surface and groundwater extraction to satisfy the rising demand of fresh water, some changes must be done. Alternative water management approaches, such as water reuse strategies, are needed to satisfy further increases of potable water demand (PNUEA, 2001). By reducing potable water consumption in urban areas there are obviously important economical savings that can be achieved, not only for the population but also for governments. The inefficient use of water has negative consequences for the environment and represents higher energy use in treatment plants and supply systems, with extra financial and environmental costs (Santos and Taveira-Pinto, 2013).

In Portugal, agriculture and golf courses irrigation are already being supplied by treated wastewater, mainly in the south of the country (Angelakis et al., 1999). These reuse schemes apply the wastewater effluent after being treated in a centralized wastewater treatment plant. This practice gains major importance in countries with severe water stress problems and droughts. For this reason, governments like the Australian are developing policies and on-ground actions for conserving and recycling water (Pinto et al., 2010).

Greywater reuse is a process with great potential to achieve sustainable water consumption in urban areas and provide potable water savings once it promotes a second use of water, thus reducing potable demand and the amount of wastewater produced in buildings. If applied in a large scale, this practice will also reduce the amount of wastewater produced and minimize the negative impacts and costs of water extraction and wastewater treatment.

GREYWATER CHARACTERIZATION

Wastewater produced in buildings is often separated, depending on its characteristics and production places, into greywater and black water. Greywater is defined as wastewater from showers, baths, washbasins, laundry, washing machines and kitchen sinks. However, many authors and guidelines around the world consider greywater as wastewater from showers, baths and hand basins only, excluding the more contaminated water from washing machines and kitchen sinks. Compared to domestic wastewater, greywater generally contains less organic pollutants, less nutrients but a high amount of tensides (Paris and Schlapp, 2010). However, its reuse involves risks as it can contain several pollutants, depending on its source, such as organic matter, suspended solids, nutrients, detergents, faecal microorganisms including pathogens such as Salmonella (Burrows et al., 1991; Rose et al., 1991; Christova-Boal et al., 1996; Surendan and Wheatley, 1998; Almeida et al., 1999; Dixon et al., 1999; Nolde, 1999; Casanova et al., 2001; Eriksson et al., 2002; Friedler, 2004; Friedler et al., 2005; Friedler et al., 2006a; Gilboa and Friedler, 2008).

Friedler et al., (2013) further divided greywater into: *light greywater* (from bathtubs, showers and bathroom washbasins) and *dark greywater* (from kitchen sink, dishwasher and washing machine). The quality characteristics of greywater (light and dark) are highly dependent on the behavior of the dwellers, the appliances connected and the chemicals used (Friedler et al., 2013). When greywater is considered for on-site reuse, it is advisable to identify the proportional volume and pollutant contribution of each generating appliance and select, according to the reuse needs, the streams with the lowest pollutant load (Friedler et al., 2013). In a perspective of greywater reuse and when the balance between demand and production is positive, it is common to reuse only light greywater (Jefferson et al. 2004), which is expected to be less polluted and easier to treat (Godfrey et al. 2009). Summarizing, it can be said that the volume and concentration of this separately collected wastewater flow depend on the consumer behavior and vary according to its source (Paris and Schlapp, 2010).

In terms of flow patterns, greywater varies significantly in diverse buildings, due to different water uses and human behaviors. Average values are presented by Friedler et al. (2013) based on wastewater discharges from a large number of individual households: the presented characteristic greywater flow patterns (which were similar to backwater) exhibit a significant morning peak at 6:00-10:00 followed by smaller evening peaks (19:00-23:00). Peak intensity and timing varies from one place to another with differences explained by different lifestyles and cultures. It is also important to notice that greywater production also varies from week days to weekend days as many people tend to spend more time at home during the weekend.

A correct characterization of greywater, prior to installing a reuse system, is very important to assess its potential for a direct reuse or, if not possible, the correct definition of the treatment system in order to get a feasible, cost-effective one. The main aspects for a correct design of the treatment unit are the site conditions and greywater characteristics which are quite variable among households (Mandal et al. 2011). This variability is mainly due to the type of detergents used, washed products, lifestyle of occupants and other practices followed at household levels (Pinto et al., 2010). When analyzing different types of buildings, other than houses, this variation can be more significant.

Domestic Greywater

This section describes a qualitative characterization of total and light greywater produced in households as well as a qualitative characterization of greywater by domestic devices. Greywater quality is expected to vary widely due to different uses given to sanitary installations (in this case, showers and washbasins). In public buildings this variation should be more significant once those installations are used by diverse people and, consequently, the amount of dirtiness, soaps, bathing products and other pollutants varies accordingly with every usage.

Although conceived to be clean, this water is polluted and contaminated. Greywater contributes significantly to wastewaters parameters such as biochemical oxygen demand (BOD), chemical oxygen demand (COD), total suspended solids (TSS), ammonium (NH4+), total phosphorous, boron, metals, salts, surfactants, synthetic chemicals, oils and greases, xenobiotic substances, and microorganisms (Friedler, 2004; Wiel-Shafran et al., 2006; Travis et al., 2008; Eriksson et al., 2002; Gross et al., 2007; Eriksson and Donner, 2009). Untreated domestic wastewater typically contains 50 to 100 mg/L of oils and greases with approximately 2/3 of the load contributed by greywater (Gray and Becker, 2002; Tchobanoglous et al., 2003). All of these components have potential negative environmental and health impacts.

Matos et al., (2012), in order to characterize total greywater (TGW) produced in households changed the drainage system of a dwelling located in Quinta da Casa Nova in Sabrosa, Vila Real District, in Tras-os-Montes and Alto Douro region, northern Portugal. For that purpose, greywater from a bathroom (including bathtub, toilet and bidet) from the kitchen (constituted by the kitchen sink and dishwasher) and from the laundry (draining the water generated by the washing machine) was collect. The daily occupancy of the house was 4 to 6 people. This greywater was sent to a stainless steel AISI 316L tank with 318 L capacity. The storage volume was defined to store all the greywater generated during a day, ensuring thus the homogenization of water from various appliances. Additionally, to characterize light greywater (LGW) produced in the dwelling, the drainage of water from the kitchen and laundry was disconnected from the system described above. Thus, only the greywater from the bathtub and washbasin was drained into the storage tank.

The most remarkable mark of these waters is the great qualitative variability, which persists even with a high number of collected samples (Friedler and Butler, 1996). In the study presented by Matos et al. (2012), and in agreement with other precedent, very different values for most parameters were verified, especially mean concentrations of dissolved oxygen, total coliforms and faecal coliforms. LGW presented large amounts of organic matter and are heavily contaminated (values greater than 104 CFU/100 mL). In general and as it would be expected, the concentration of the analyzed parameters in the TGW is superior to the LGW (Figure 1).

There are, however, some exceptions such as copper (Cu), iron (Fe), zinc Zn) and total organic carbon (TOC), where the concentration is greater in the LGW. For the microbiological parameters, aluminium, cadmium, phosphorus and sodium, TSS, sulphates, COD and RAS concentration difference between the LGW and TGW is evident, and is significantly higher in TGW.

Values of the analyzed parameters are highly variable depending on several factors: consumers behaviors and activities, the amount of soaps and bathing products used, washing temperatures, etc.

However, it is most evident pollutant and contaminant load in TGW than in LGW, in particular at the microbiological level, and, in principle, it is easier to treat LGW in order to obtain an effluent for reuse. These finding are in agreement with other referenced work (Almeida et al., 1999; Butler, 1991; Butler et al., 1995).

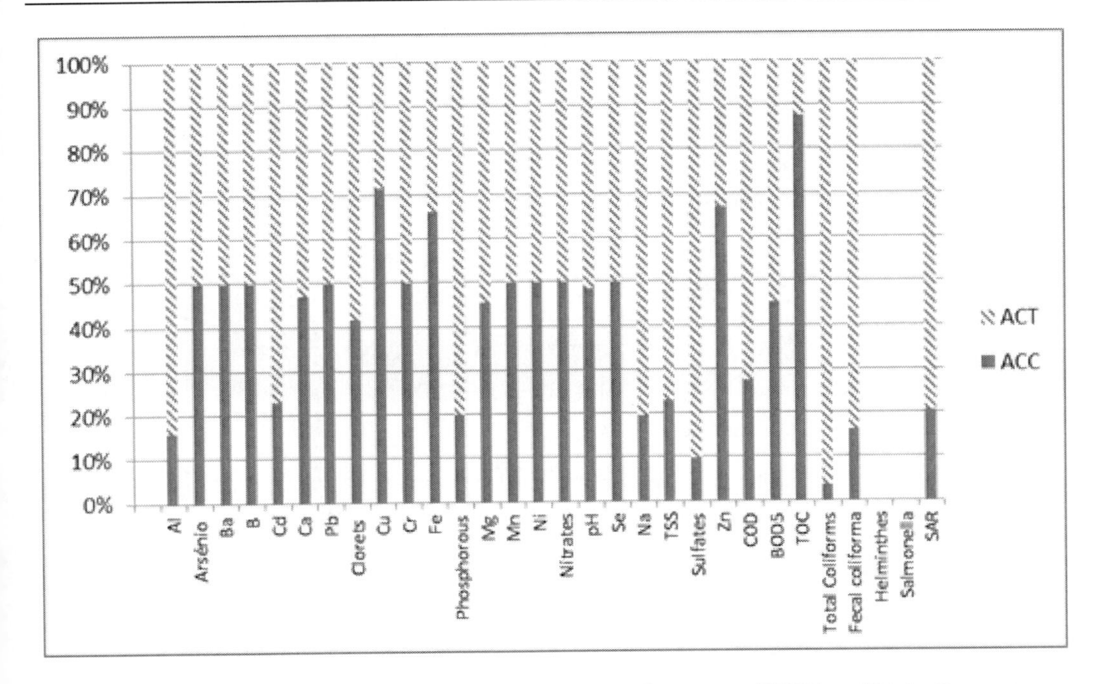

Figure 1. Relative concentrations of each parameter in Total Greywater (TGW) and Light Greywater (LGW) (Matos el at., 2012).

Greywater Quality per Domestic Device

There is some research already made about the quality of greywater and its variation by source (Tables 1 and 2) and within source type (Matos, 2009; Matos et al., 2010). For instance, literature reports important differences for washing machines between the effluents of different cycles and the same can be expected from dishwashers (Rose et al., 1991; Burrows et al., 1991; Christova-Boal et al., 1996; Surendanand Wheatley, 1998; Shin et al., 1998; Nolde, 1999; Eriksson et al., 2002; Friedler, 2004). Laundry greywater exhibited a high range of the values of suspended solids, salts, nutrients, organic matter and pathogens which arise from washing of clothes using detergents (Christova-Boal et al., 1996). In fact, some activities such as washing faecal contaminated laundry, childcare and showering add faecal contamination to greywater (Ottoson and Stenström, 2003). Occasionally, gastrointestinal bacteria such as Salmonella and Campylobacter can be introduced by food-handling in the kitchen (Cogan et al., 1999). Greywater may have an elevated load of easily degraded organic material, which may favour growth of enteric bacteria such as faecal indicators and such growth has been reported in wastewater systems (Manville et al., 2001). Kitchen greywater is reported as the highest contributor of oils and greases in domestic greywater, but oils and greases are present in all greywater streams (Friedler, 2004).

As demonstrated, the chemical, physical and microbiological characteristics of greywater are quite inconstant among households due to the type of detergents used, type of products being washed, life style of occupants and other practice followed at household levels. In order to characterize greywater quality per domestic device, Matos et al., (2012) took independent samples from eight distinct houses collected and treated at the same day.

Table 1. Values for physical-chemical parameters and nutrients in greywater

Parameters	WC	Bathtub	Washbasin	Kitchen sink	Washing machine	Dishwasher
Turbidity (NTU)	60-240[1]	92[2] 28-96[3] 49-69[4]	102*[2]		50-210[1] 108[2] 14-296[3]	
Total solids (mg/L)		631[2] 777-1090[5] 250[6]	558[2] 835[5]	1272-2410[5] 2410[6]	658[2] 350-2091[5] 410-1340[6]	45-2810[5] 1500[6]
Total volatile solids, TVS (mg/L)		318[2] 533[5] 190[6]	240[2] 316*[5]	661-720[5] 1710[6]	330[2] 125-765[5] 180-520[6]	30-1045[5] 870[6]
Total suspended solids, TSS (mg/L)	48-120[1]	76[2] 54-303[5] 120[6] 54-200[7]	40[2] 259[5] 181[7]	625-720[5] 720[6] 185[8]	88-250[1] 68[2] 65-280[5] 120-280[6] 165[7]	15-525[5] 440[6] 235[7]
Volatile suspended solids, VSS (mg/L)		102[5] 85[6] 9-153[7]	36-86[5] 72[7]	459-670[5] 670[6]	97-106[5] 69-170[6] 97[7]	10-424[5] 370[6]
Chemical oxygen demand, COD (mg/L) *Total*	210-230[5]	645[5] 210-501[7] 100-633[9] 282[10]	95-386[5] 298[7] 383[10]	936[2] 644-1340[5] 1079[7] 1380[10] 15-26[11]	725[2,10] 1339[5] 1815[7]	1296*[5]
Dissolved	165[5]	319[5] 184-221[7]	270[5] 221[7]	679*[5] 644[7]	996[5] 1164[7]	547*[5]
Biochemical oxygen demand, BOD_5 (mg/L) *Total*	173[5]	216[2] 170[6] 424[5] 192[10]	252[2] 33-236[5] 236[10]	536[2] 1460[6] 530-1450[5] 5[8] 2762[10]	48-290[1] 472[2] 280-470[5] 150-380[6] 282 [10]	390-699*[5] 1040[5]
Dissolved	76-200[1] 75[5]	237[5]	93*[5]	377*[5]	48-290[1] 381*[5]	262*[5]
Total organic carbon, TOC (mg/L) *Total*	91[5]	104[2] 30-120[5] 100[6]	40[2] 119[5]	582*[5] 880[6]	381[5] 100-280[6]	234*[5] 600[6]
Dissolved	47[1]	59[5]	74*[5]	316*[5]	281*[5]	150*[5]
pH	6.4-8.1[1] 7.1[5]	7.6[2] 6.7-7.4[4,5]	8.1[2] 7.0-8.1[5]	6.5[5] 6.3-7.4[8]	9.3-10.0[1] 8.1[2] 7.5-10.0[5]	9.3-10.0[1]
Total Nitrogen (mg/L)	4.6-20[1]	17[6] 5-10[9]		74[6] 15.4-42.5[8]	1-40[1] 6-21[6]	40[6]
Ammonia	<0.1-15[1] <0.9-1.1[5]	1.6[2] 0.1-0.4[3] 1.2[5] 2[6] 1.1-1.2[7] 1.3[10]	0.5[2] 0.4-1.2[5] 0.3[7] 1.2[10]	4.6[2] 0.6-6.0[5] 6.0[6] 0.3[7] 0.2-23.0[8] 5.4[10]	<0.1-0.9[1] 10.7[2] 0.06-3.5[3] 4.9-11.0[5] 0.4-0.7[6] 2.0 [7] 11.3[10]	4.5-5.4[5] 4.5[6]

Parameters	WC	Bathtub	Washbasin	Kitchen sink	Washing machine	Dishwasher
Nitrates and Nitrites	<0.05-0.2[(1)]	0.9[(2)] 0.4[(6)] 4.2-6.3[(7)] 0.4[(10)]	0.3[(2)] 6.0[(7)] 0.3[(10)]	0.5[(2)] 5.8[(3)] 0.3[(6)] 0.6[(10)]	0.1-0.3[(1)] 1.6[(2)] 0.4-0.6[(6)] 2.0[(7)] 1.3[(10)]	0.3[(6)]
Total Phosphorus (mg/L)	0.11-1.8[(1)]	2.0[(6)] 0.2-0.6[(10)]		74.0[(6)]	0.06-42[(1)] 21-57[(6)]	68[(6)]
Phosphates	4.6-5.3[(5)]	1.6[(2)] 10-19[(5)] 1.0[(6)] 5.3-19.2[(7)] 0.9[(10)]	45.5[(2)] 13-49[(5)] 13.3[(7)] 48.8[(10)]	15.6[(2)] 13-31[(5)] 31.0[(6)] 26.0[(7)] 0.4-4.7[(8)] 12.7[(10)]	101.0[(2)] 4-170[(5)] 4-15[(6)] 21.0[(7)] 171.0[(10)]	32-537[(5)] 32.0[(6)]

*Mean of 150 samples; [(1)] Christova-Boal *et al.* (1996); [(2)] Surendran and Wheatley (1998); [(3)] Rose *et al.* (1991); [(4)] Burrows et al. (1991); [(5)] Friedler (2004); [(6)] Siegrist *et al.* (1976); [(7)] Almeida *et al.* (1999); [(8)] Shin *et al.* (1998); [(9)] Nolde (1999); [(10)] Laak (1974).

Table 2. Range values for microbial parameters analysed in greywater obtained in kitchen, laundry and bathroom

Parameters	Kitchen	Laundry	Bathroom
Escherichia coli (number/100 mL)	$1.3 \times 10^5 - 2.5 \times 10^8$[(1)]		
Thermotolerant *E. coli* (MPN)	$9.4 \times 10^4 - 3.8 \times 10^8$[(1)]		
Faecal Streptococcus (MPN)	$5.1 \times 10^3 - 5.5 \times 10^8$[(1)]	$23 - < 2.4 \times 10^3$[(2)] $1 - 1.3 \times 10^6$[(3)]	$79 - 2.4 \times 10^3$[(2)] $1.0 \times 10^4 - 7.0 \times 10^6$[(3)]
Total Coliforms (MPN)	$6.0 \times 10^4 - 4.0 \times 10^7$[(1)]	$2.3 \times 10^3 - 3.3 \times 10^5$[(2)] $8.5 \times 10^5 - 8.9 \times 10^5$[(3)] 7.0×10^5[(4)]	$500 - 2.4 \times 10^7$[(2)] $8.2 \times 10^3 - 7.0 \times 10^4$[(3)] $5.0 \times 10^4 - 6.0 \times 10^6$[(4)]
Faecal Coliforms (MPN)		$110 - 1.09 \times 10^3$[(2)]	$170 - 3.3 \times 10^3$[(2)] $1.0 \times 10^3 - 2.5 \times 10^3$[(3)] $6.0 \times 10^2 - 3.2 \times 10^3$[(4)]

[(1)] Günther (2000); [(2)] Christova-Boal et al. (1996); [(3)] Siegrist et al. (1976); [(4)] Surendran and Wheatley (1998). MPN: most probable number.

The houses were single-family houses, varying in the number of inhabitants from 2 to 6 per house. Greywater was separated by its origin and water samples in both rooms that generated effluents: kitchen and bathroom were collected. In each room, samples were collected in each production place:

- kitchen: sink, dishwasher and washing machine;
- bathroom: washbasin, bathtub and bidet (this last appliance is widespread in the Mediterranean region).

Samples from raw greywater were analyzed for pH, conductivity, TDS and COD. To investigate the concentration of bacteria in raw greywater, total and faecal coliforms were also counted.

Comparing the mean values of pH recorded for drinking water of different houses with greywater from different sources, it appears that, except for the greywater that came from the bathtub and kitchen sink, this value is higher in greywater. Washing machines and dishwashers revealed the highest values with respect to conductivity (Figure 2). In fact, the water from the dishwasher had values 20 times higher than the drinking water and water from the washing machine, 50 times higher. The results for this parameter lead to very high SDT values especially in these two domestic devices. The COD values were high, with the exception of water from the bidet, reaching a maximum of 1781.5 mg/L in the sink. Most of the COD derived from the chemicals used and is therefore higher in the laundry and kitchen, with great variations from house to house. Microbiological contamination of total and faecal coliforms was always very significant, except for washing machine that did not show presence of any faecal coliforms, whatever the dilution used. The domestic devices from kitchen and laundry were the main pollutant concentration producers, although the bath also contained significant amounts of faecal coliform. This reveals that greywater from the kitchen may contain numerous microorganisms from the food washing and is usually the most polluted source.

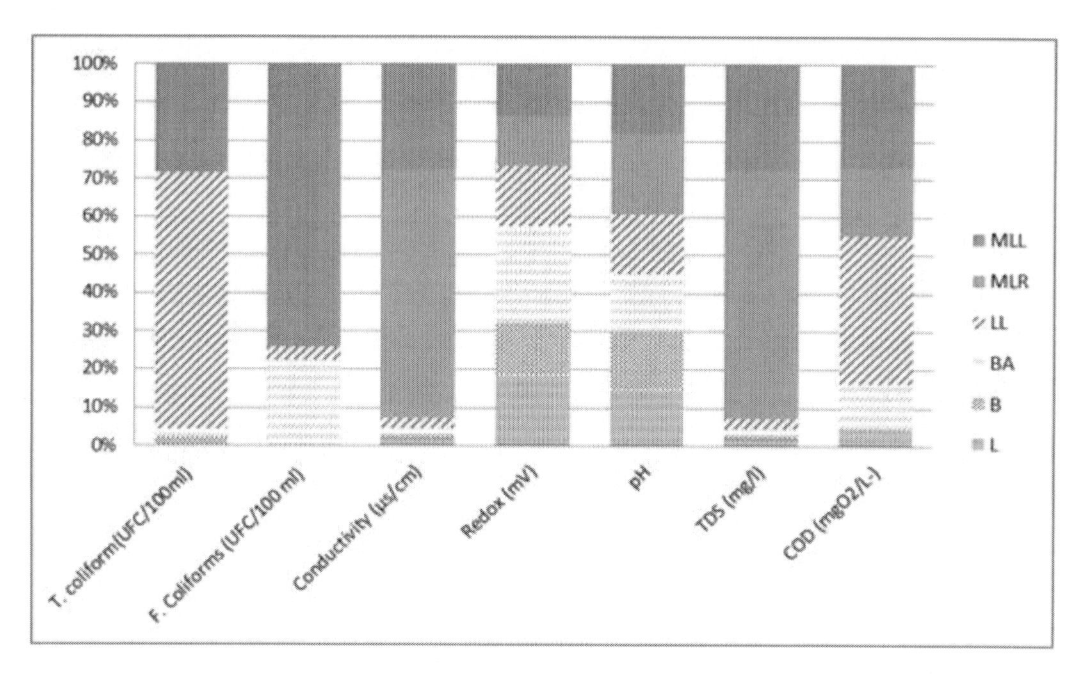

Figure 2. Relative concentrations of each parameter in each domestic device. MLL- Dish Washer; MLR- Washing machine; LL- Kitchen sink; BA- Bathtub; L- Washbasin; (Matos et al., 2012).

Greywater from Public Buildings

As referred before, greywater quality is expected to vary widely due to different uses given to sanitary installations. In public buildings this variation should be more significant once those installations are used by diverse people and, consequently, the amount of dirtiness, soaps, bathing products and other pollutants varies accordingly with every usage.

A study made by Santos (2012) to characterize greywater from public buildings, included the collection of water produced in wash basins of both a changing room exclusively used by the employees and a public toilet in a restaurant, both located in the Faculty of Engineering of Porto University, in Portugal. Other samples of greywater were also collected from showers of a changing room for the employees of a wastewater treatment plant in Vila Nova de Gaia, Portugal. The process of collection and treatment of greywater was repeated four to five times in each location, on a weekly frequency. Samples of raw and treated greywater were taken, preserved and analyzed in accordance with Standard Methods for the Examination of Water and Wastewater (AWWA-APHE-WPCF 1989). Parameters measured include pH, Total and Volatile Suspended Solids (TSS and VSS), Total Solids (TS), Ammonia (NH_3), Total Phosphorus (TP), and Organic Matter (COD and BOD). Microbial parameters were also analyzed, namely total and fecal coliforms, according to Membrane Filter Technique (Clesceri et al. 1989), in the Microbiological Laboratory of the Faculty of Engineering, University of Porto.

Since domestic wastewater is generated as by-product of human activities, its quality, in terms of constituents and concentrations, is expected to vary widely. Greywater quality is no exception as it was shown in the previous section and can also be seen by the wide range of values for each parameter presented in Table 3. In public buildings this variability can be explained by the different uses given to sanitary installations (showers and washing basins) by diverse people and consequently the amount of dirtiness, soaps, bathing products and other pollutants varies accordingly with every usage.

High concentrations of solids (TSS, VSS and TS) were registered in samples from washing basins of the employees' changing room. This can be explained by the activities of personal hygiene carried out in that changing room: showering shaving, combing and tooth brushing, especially in the morning. The washing basins of the restaurant are used mainly for hand washing. Higher concentrations of ammonia and BOD were registered in samples from shower effluents, probably due to the existence of urine. The same pattern was observed in total phosphorus which might be caused by the amount and variety of hygiene products used. COD concentrations, ranging from 158 to 263 mg/l, were similar among samples from the three collection places, and were also comparable to the ranges of value presented in previous studies. (Al-Jayyousi 2003; Friedler et. al, 2006; March et. al, 2004; Friedler et. al, 2005; Friedler and Alfiya, 2010).

Table 3. Greywater from public buildings: quality parameters

PARAMETER	Washing basins (changing room)			Washing basins (restaurant toilet)			Showers (changing room)		
	Min	Max	Mean	Min	Max	Mean	Min	Max	Mean
pH	7,1	7,5	7,3	7,0	7,4	7,2	6,6	7,1	6,9
TSS [mg/l]	45	184	83	44	68	55	44	71	58
VSS [mg/l]	40	141	71	47	60	51	41	46	43
TS [mg/l]	278	670	378	244	380	307	208	622	326
NH3 [mg/l]	<1	2,2	<1	<1	1,1	<1	5,8	13,6	8,6
TP [mg/l]	0,3	1,3	0,6	0,5	1,0	0,8	0,9	1,7	1,3
COD [mg/l]	76	287	166	162	285	209	167	247	197
BOD [mg/l]	33	89	57	38	85	65	41	296	129

PUBLIC PERCEPTION TOWARDS GREYWATER REUSE

Any strategy of water reuse that involves changes in people's habits must achieve social acceptance to be successful (Friedler et al., 2006). Water reuse needs to include community and stakeholder participation from the beginning so public acceptance has to be assessed. In order to evaluate public acceptance of these practices there are three categories of surveys described in the literature. The first, attempts to establish the general attitude toward water reuse by asking the public a wide range of questions. The second category seeks public opinion on forthcoming water reuse projects. The third, examines public attitude in places where reuse schemes have already been put in place.

Most of the studies belonging to the first category concluded that a large majority of the public supports the concept of water reuse, although this acceptance is reduced when the degree of contact of people with the reclaimed water increases (Bruvold, 1984; Denlay and Dowsett, 1994; Jeffrey and Jefferson, 2003; Crook, 2003). Reasons like water conservation, environmental issues, health issues and costs of treatment and distribution of water were outlined as justification for support or objection options (Bruvold, 1988). Studies included in the second category reveal general supportive opinions on water reuse for toilet flushing, clothes washing and garden irrigation (Van der Hoek et al., 1999; Marks et al., 2003). Studies that examine public attitude in places where reuse schemes have already been put in place (third category) found that cost savings was the most important reason to support water reuse in irrigation, car washing, toilet flushing. This was followed by the positive effects on the environment and saving scarce potable water sources (Marks et al., 2003).

Much of the research about water reuse acceptance conducted during the 1970s and 1980s in USA has been summarized by Bruvold (1988). These studies indicated 90% support of wastewater reuse in recreational parks, golf courses, lawns, gardens and hay pastures irrigation. 80-90% support was often reported for wastewater reuse in irrigation of dairy pastures, orchard, vineyard and vegetable crops. 70-90% support was indicated for household toilet flushing and clothes washing. 60-75% support was reported for reuses like swimming and bathing at home that correspond to high contact options. 30-60% was consistently reported for reuses that involve direct human ingestion, like drinking and cooking.

In a recent study, Kantanoleon et al., (2007) described results of a survey conducted in Chalkida (Greece), a Mediterranean city, where 76% of the population surveyed supported wastewater reuse in industrial applications. However, as in other studies (Bruvold et al., 1981; Bruvold, 1984; Denlay and Dowsett, 1994; Jeffrey and Jefferson, 2003; Crook, 2003), the opposition to specific reuse option increased with the degree of contact, for example 69% did not support wastewater reuse in playgrounds irrigation, 80% did not support the use of wastewater in animal crops irrigation, while 94% did not support potable reuse.

Friedler et al., (2006) conducted a survey in Haifa (Israel) in order to determine the attitude of a sample of the Israeli urban public towards various water reuse options. The survey clustered the reuse options into three reuse categories, namely: low, medium, and high contact levels. The study found that a high proportion of the participants supported medium contact reuse options such as sidewalk landscaping (95%), domestic WC flushing (85%) and firefighting (96%). Higher contact reuse options such as domestic laundry (38%), preserved food industry (13%), and recharge of potable aquifer (11%) found much lesser support. Support for low contact reuse options was lower than expected with 86% for field crop irrigation, 62% for aquifer recharge for agricultural irrigation, and as low as 49% for orchard irrigation. In other studies (Bruvold 1984; EPA, 1992; Crook et al 1994; Hartley, 2006), high support was given by the participants to the low and medium contact reuse options. According to Friedler et al., (2006) it can be asserted that it is safe to say that the majority of water sector professionals in arid and semi-arid regions admit reusing wastewater effluent in non-potable end-uses, however this cannot be assumed for the public in general. This is related in some cases with insufficient and/or inappropriate dissemination of information to the public and in other cases with a lack of trust in centralized organizations (Jeffrey and Temple, 1999). In fact, a different study reveals that people who attended the workshops and activities disseminating information on wastewater reuse supported a wider range of reuse options that those who had not (Simpson, 1999).

There are also few studies that tried to characterize the typical objector to water reuse in potable reuse in terms of age, gender, socioeconomic status and level of education (Bruvold, 1984; Marks, 2004). Bruvold described the characteristic objector to potable water reuse as having a low socioeconomic status, being older and having low awareness of water and environmental issues. As in Bruvold (1984), Marks (2006) reports that in some surveys females were less supportive than males. Actually, several studies found that age (Stone and Kahle, 1974; Lohman and Milliken, 1985; McKay and Hurlimann, 2003; Hurlimann 2007a; Dolnicar and Schäfer 2009) and gender (Baumann and Kasperson, 1974; Lohman and Milliken, 1985; Tsagarakis et al. 2007; Hurlimann, 2007a; Nancarrow et al. ,2008; Dolnicar and Schäfer, 2009) do have an influence on the level of support of recycled water projects. Marks and others also found that higher education tends to be associated with higher support to water reuse options (Bruvold, 1972; Stone and Kahle, 1974; Flack and Greenberg, 1987; Lohman and Milliken, 1985; Alhumoud et al., 2003; Menegaki et al., 2006; Hurlimann, 2007a; Dolnicar and Schäfer, 2009; Robinson et al., 2005). Marks noted that freelance, professional and white collar workers were more receptive to non-potable reuse options. On the other hand recent studies, who examined non-potable reuse (Jeffrey and Jefferson, 2003; Friedler et al., 2006a), found no correlation between level of support and age and gender. In Friedler et al.'s study (2006a) no correlation was found between any demographic characteristic examined and support for medium contact options.

For medium contact options, Friedler et. al.'s study revealed that perceived financial gain (individual and/or communal) and positive public opinion enhances support, while perceived health effects negatively affects the degree of support. Other studies reported that health concern and consequently risk perception, negatively influences attitudes through water reuse projects (Olson et al., 1979; Dishman et al., 1989; Po et al., 2005; Marks et al., 2006; Baggett et al., 2006; Hurlimann, 2008; Hurlimann et. al., 2008). Trust in authorities and awareness of water and environmental issues did not have a significant effect on support for medium contact reuse options in Friedler's study. Participants in the survey who identified themselves as supporters of wastewater reuse revealed that the most important reason for their support was "water saving", followed by "minimization of importing water from abroad" while "environmental improvement" ranked as the third most frequent response together with "infrastructure cost saving". On the contrary, in a research carried out in Australia, a conjoint analysis was used to evaluate participant's preferences for various attributes of recycled water (color, odor, salt content) for various uses and these were found to be prominent reasons to the level of support (Hurlimann and MacKay, 2007). Also Domènech and Saurí, (2010), found out that the perception of health risks, operation regimes, perceived costs and environmental awareness are, in different degrees, significant determinants of public acceptance. These authors concluded that improving the level of knowledge of these systems among users would reduce the risk of social refusal of the new technology. Public authorities and suppliers need to stimulate social learning processes with specific actions, and build trust among residents in the new governance network if decentralized and alternative water supply systems are to find a place in the everyday life of urban populations. In other studies a clear correlation was found between economic gain and trust in authorities and support for water reuse (Lohman and Milliken, 1985; Jeffrey and Jefferson, 2003; Marks et al., 2003; Hurlimann and McKay, 2004; Friedler et al., 2006; Hurlimann, 2007b; Hurlimann, 2007c). Hurlimann and McKay, (2004) found that the degree of trust that an individual has for a water authority is proportionate to the individual's level of confidence that supply of reused water would not pose risks to their health or garden.

Matos et al. (2013) presented the results of a multiple choice survey that attempted to establish the general attitude towards water reuse by asking academics in University of Trás-os-Montes e Alto Douro (Portugal) a wide range of questions. The survey included 20 reuse options, which were clustered into three reuse categories, namely: low, medium, and high contact levels. Correlation analysis between the support level of low, medium and high contact options and demographic characteristics, personal and environmental beliefs was performed. Results showed that a high proportion of the participants supported low and medium contact reuse options. Correlation between the income classes and to the support level of medium and high reuse options and between education level and the support for high contact reuse options was confirmed. Also Muthukumaran et al. (2011) concluded that community receptivity for reusing greywater is highest for uses such as irrigation and flushing toilets but decreases if personal contact with greywater is needed.

Community acceptance of water reuse may be significantly influenced by regional circumstance (Kahn and Gerrard, 2006). In particular severe shortage of freshwater supplies is likely to encourage communities to look for alternative sources. Windhoek (Namibia), for example, suffers a combination of very low rainfall, high evaporation and limited catchment area, and so, the city now recycles water from sewage treatment plants directly to drinking water plants, to supply about one third of its potable requirements (Khan and Gerrard, 2006). Another example is Singapore, a small island with extremely limited natural fresh water supplies, being heavily dependent on Malaysia, for much of its potable water. In 2000, Singapore commissioned its first NEWater advanced wastewater reclamation plant to supply potable reuse, today, 10 years later and with the 4th NEWater treatment plant commissioned in 2010, NEWater initiative supplies 30% of its national water demand. The lesson that can be learnt from Singapore's case is that prospect of secure, self sufficient water supplies combined with trust in authorities lead to high levels of support of water reuse (Macpherson and Law, 2003).

It is manifested from the above discussion that acceptance of water reuse schemes in particular communities varies over time and location. Therefore, ongoing studies at the local area are always necessary to keep pace with community sentiment in each instance. In fact, whenever exists varying conditions of water availability, climate, culture, socio-economical background there is the implicit need to gather robust data, and so, these is the main motivation to the present research.

GREYWATER REUSE SYSTEMS

A detailed study and correct definition of a greywater reuse system is crucial to get a feasible, cost-effective one and take the best benefits from it. As explained before, this is a practice with significant potential for water savings in urban areas but needs to be correctly designed once local conditions and greywater characteristics are quite variable among households (Mandal et al. 2011).

Types of Greywater Reuse Systems

The reutilization of greywater demands an installation to collect greywater from production places, drain it to a storage tank, treat it according to the quality needed in the intended reuse places and a distribution system, including pumps if necessary. There are several configurations for these systems. The referred components are the same amongst them, except for the treatment unit which depends on the required treatment for each case.

The most simple greywater reuse systems, don't require any kind of treatment once they make a *direct reuse.* Greywater produced in washbasins and showers is directly reused in toilets or garden irrigation with no, or minimum, storage (Figure 3) which means that greywater must be reused as soon as it has cooled. The main advantages of these systems are their small cost and the fact that there is no need to install large equipment, providing an easy and inexpensive way to save water in buildings. If an irrigation system is supplied by these type of systems, its configuration and the type of watered crops must be in accordance to normative documents that may possibly exist in each location (such as the Portuguese standard for water reuse in irrigation (IPQ, 2005)). In the UK, for example, it is recommended to use it only in sub-surface irrigation and non-spray applications (BSI, 2010).

Figure 3. Direct reuse systems (Source: www.pt.roca.com, last accessed on 27.04.2013).

More complex systems, with storage tanks, must have a treatment unit once untreated greywater deteriorates rapidly in storage due to temperature and organic matter in greywater. This warm, nutrient-rich water provides ideal conditions for bacteria to multiply, resulting in odor problems and degradation of water quality. Greywater may also have harmful bacteria, which could present a health risk without adequate water treatment on inappropriate use (EA, 2011). Greywater can be contaminated by activities such as bathing and clothes washing. Disease-causing organisms in greywater are principally transmitted through ingestion of greywater via contaminated hands, aerosols from spray irrigation (usually only allowed for use with suitably treated greywater), or indirectly through contact with contaminated items such as grass, soil, toys and garden implements (Standards Australia, 2001).

Greywater reuse systems whit *short retention* usually provide a very simple treatment: only a coarse filtration prior to the storage tank, sedimentation within it and skimming debris off the surface. As explained above, greywater must not be stored for too long and must be replaced by potable water if it is not used in a short space of time. This means that water savings provided by these systems are depending to usage patterns. But it is important to highlight their low installation and maintenance costs (EA, 2011; BSI, 2010).

Basic *physical/chemical* reuse systems are similar to the prior ones but include chemical disinfection to prevent bacterial growth (Figure 4) but the choice of chemicals disinfectants must be done carefully to prevent damages in the consumption places especially if the treated water is used in irrigation (EA, 2011). Chlorinated disinfectants are highly effective but their use is not friendly from the environmental point of view, once it can result in the production of undesirable disinfection byproducts, like chloramines, trihalomethanes and haloacetic acids, harmful to human health (Metcalf and Eddy. et al. 2003; Al-Jayyousi, 2003; Gual *et. al*, 2008). Disinfection by chemicals can be replaced by UV lamps to minimize these problems although, to make it effective, a correct filtration must be done to prevent microorganisms to hide from UV light, behind the solids. These systems require a careful installation and easy access for maintenance. Once storage retention time gets longer, water savings will probably be higher than short retention systems and can go up to 32% of total water use (EA, 2011).

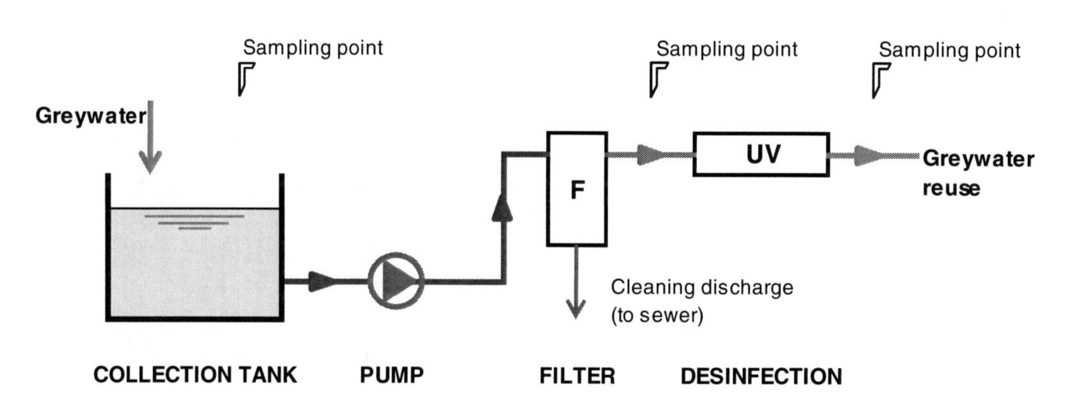

Figure 4. Basic physical/chemical reuse systems presented by Santos et al. (2011).

A more complex treatment is made by *biological systems* that use bacteria to remove organic matter from greywater. These systems demand an introduction of oxygen in the water for the organisms to make the purification process. This aeration of greywater, usually made by aerators or compressed air systems, demands exploration costs that can be minimized if the oxygen is provided by aquatic plants (reeds, for example) that introduce oxygen in the water by the roots. Greywater can pass through the soil/gravel in which the reeds are growing and the bacteria fed by the oxygen decompose the organic matter (EA, 2011). Reed beds and similar installations require, however, a large outdoor space.

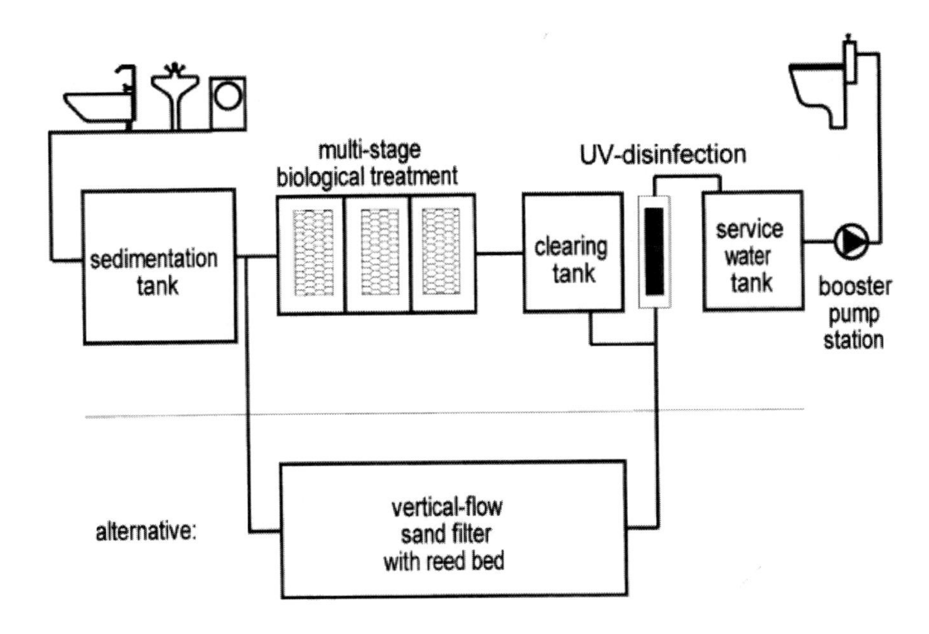

Figure 5. Biological treatment in the greywater reuse systems presented by Nolde (1999).

The most complex systems to reuse greywater are *bio-mechanical systems* that combine physical, biological and disinfection processes. Greywater is filtrated and sent to a biological reactor where bacteria are encouraged to digest organic matter by oxygen introduced in the reactor. Treated greywater is then disinfected by introducing chlorine or UV light. An example of these systems, already implemented in many buildings is Pontos Aquacycle, from Hansgrohe (Figure 6). Membrane ultrafiltration may also be used in this kind of greywater reuse systems. The combination of physical and biological treatment generally produces the highest quality water but it also uses a significant amount of energy, is expensive to purchase and operate and maintenance costs are uncertain.

Figure 6. Pontos Aquacycle (source: www.hansgrohe.com, last accessed on 01-05-2013).

The choice of which type of system is appropriate for a certain building depends mainly on the water quality desired and the available space to install its components. Friedler et al. (2006) presented three pilot-scale treatment units of light domestic greywater from a residential building. The simplest one was a stand-alone filtration with a gravity sand filter followed by chlorine disinfection, which was the less effective promoting a reduction on pollutant parameters that was not enough to accomplish local guidelines for toilet greywater reuse. This unit produced treated greywater with average values for SS, COD and BOD of 32, 130 and 62 mg/L, corresponding to a reduction of 65, 38 and 10% respectively. The other two treatment units studied were more complex, including a rotating biological contractor and membrane bioreactor, and produced water of "excellent quality" according to local guidelines.

Another study presented by Gual et al. (2008), showed a chlorination and filtration unit for greywater produced on the bathrooms of an hotel. This system provided a significant treatment and produced effluents having average SS and COD of 27 and 55 mg/l, respectively. In Jordan, Al-Jayyousi (2003) presented a simple sand filtration unit to treat and recover greywater for irrigation. Average values of SS and BOD in treated greywater were, in this case, 189 and 392 mg/l revealing a reduction of 40 and 74%, respectively. In their study, Jefferson *et al.* (2004) referred that biological processes are the most suitable unit processes for treating grey water due to the characteristics of greywater. The highly variable nature of the source requires that selected technologies must be inherently robust in their operation. One potential area of concern is the high COD/BOD ratio and nutrient deficiency in terms of both macro and micro nutrients which grey water exhibits potentially retard the efficacy of biological processes.

Santos et al. (2011) studied an experimental system for greywater treatment including a collection tank, a pumping device, a filtration system and a UV disinfection unit. The system was built to provide short residence time for the collected greywater and to disinfect it immediately before its reuse, in order to minimize the possibility of the regrowth of microorganisms. The system, including the collection tank, was prepared to be easily transported and installed, requiring only electrical supply and a sewage connection to the filter cleaning discharge. Both filtration and disinfection equipment, bearing an investment of ca. 2350 € (in 2009), were chosen for easy maintenance and affordable cost. The end results of the study were relatively good, particularly related to disinfection by UV radiation. In terms of set-up, the unit is considered economic, reliable and safe to be implemented for on-site application. However, unexpectedly lower removal of SS and BOD may hinder its acceptance, as the treated water appears turbid and may cause undesirable odor after a short period of storage in toilet flushing tanks.

Integration with Rainwater Harvesting Systems

Water savings resulting from greywater reuse can be increased if rainwater is harvested and integrated in the system, providing more non-potable water to be consumed. Once usual domestic water consumption patterns indicate that greywater produced is sufficient to supply the non-potable demand, integrated systems are more feasible in large buildings where one or the other source is not sufficient to meet the intended demand on its own. These systems can either be operated as separate, independent systems or be combined into a single supply source (BSI, 2010). They must provide the necessary treatment, depending on the water quality required in non-potable purposes. However, it should be done separately, prior to the integration, as greywater demands normally a more complex treatment than rainwater (if greywater storage time is intended to be significant). The overflow of such a system must be sent to the sewer, unless the excess rainwater can be discharged prior to the point where it combines with greywater.

Both the systems and the integrated ones require a careful study to assess their feasibility, especially in large buildings, since they exert considerable investment costs and, in some cases, long payback periods. Proper analysis and design before executing these systems would have amplified the benefit and effectiveness of storage tanks (Imteaz et al., 2011).

Potential Water Savings

Greywater production and reuse potential depends, among other factors, on the type of buildings and human activities developed on it. In dwellings and residential buildings, treated greywater reuse might represent 40 to 47% reduction of total water consumption (Almeida et al. 2001). In commercial and public buildings, 48 to 53% of total water consumption is used in non-potable purposes but greywater production is smaller than the previous cases, about 2 to 15% of total water consumption (Santos 2011; Friedler and Alfiya 2010). Besides, in these buildings there are still more gains, such as administrative simplicity, association with a sustainable green image and lower specific treatment costs can be expected (Friedler and Alfiya 2010). Greywater reuse systems might also be an important input to non-potable water supply when integrated with other alternative water sources (i.e. rainwater harvesting) due to its constant availability whenever there is normal activities in the buildings.

As an example of potential economic and environmental benefits of greywater reuse, three houses with greywater systems were studied in England. The system collected greywater from the bath and bathroom washbasins and then filtered and disinfected it. Treated water was then used to flush toilets. The potable water saved ranges from 24% in a family of seven persons, until 65% in a family of three (EA, 2011). Greywater generated from a household is self-sufficient to irrigate a small garden and including the reuse in toilet flushing, about 48% fresh value can be saved (Mandal et al., 2011). In a small building with 30 inhabitants, almost 300 m^3 of potable water per year can be saved by reusing greywater (about 30% of the total water consumption (Santos, 2012)).

Previous studies show that in urban areas the most feasible greywater reuse option is for toilet flushing which can reduce individual in-house net water demand by 40-60 L/d per capita. If this reduction becomes widespread, a reduction of up to 10-25% in urban water demand can be achieved (Friedler and Hadari, 2006). In the United Kingdom, three houses with greywater reuse systems were studied. The system collected greywater from the bath and bathroom washbasins and then filtered and disinfected it. Treated water was then used to flush toilets. The potable water saved ranges from 24% in a family of seven persons, until 65% in a family of three (EA, 2011).

Investigations in Brazil determined that rainwater (RH) and greywater systems, either separated or combined, have very long payback periods, about 17 to 92 years for two studied dwellings, being the separated RH systems the more attractive ones (Ghisi, de Oliveira, 2007). For a multi-storey building, using either RH or GR separately, the payback period would be shorter than 5 years, with the GR system being a little more cost-effective than RH. Using both systems, the payback period would be longer, but still cost-effective (Ghisi and Ferreira, 2007). Chilton et al. (2000) studied a rainwater recovery system in a commercial building with large roof area and calculate a payback period of 12 years. Although the payback was longer than would normally be considered economically viable, the rainwater systems was implemented, since there are financial benefits of having reduced water and sewerage charges as well as securing an enhancement of the image as a company that is environmentally friendly.

CONCLUSION

The solution to promote a sustainable water use and guarantee its quantity and quality for future generations lies on the reduction of potable water demand by increasing the efficiency of water use. The investment in solutions and equipment to reduce potable water consumption, such as greywater reuse and rainwater harvesting systems, should be seen as a precaution for upcoming scarcity situations (EEA, 2012). This chapter showed that greywater reuse is a procedure with significant potential to reduce potable water consumption and promote a sustainable water use in urban areas. Moreover, this practice is also a great contribution to reduce the amount of wastewater produced and the consequent pollution of water resources. This benefit can be enhanced if treated greywater is used, for example, in groundwater recharge, avoiding also this amount of treated greywater to be sent to the sewer system.

Although, greywater quality is very variable, mainly due to human activities and behaviors. To define a feasible and cost-effective greywater reuse system, a correct definition of the greywater produced in the building. Another essential factor is to assess the water quality needed in the intended non potable purposes. These two information will be the base to choose the type of treatment which, as explained above, can vary widely, from simple units with no, or minimum, retention of greywater, to large complex ones that provide filtration, biological treatment and disinfection, and, in some cases, also integrated with rainwater harvesting systems. As demonstrated above, water savings provided by the implementation of greywater reuse systems in buildings can go from 10 to 65% of total water consumption in buildings, as reported by many authors, which reveal the great potential of this alternative supply to achieve significant potable water savings in urban areas.

The public acceptance about greywater reuse is, in general, quite good as demonstrated by many studies made in different countries. Although it reduces when the degree of contact of people with the reclaimed water increases revealing some prudence and mistrust about the quality of treated water produced throughout the equipment life time. Different factors such as climate conditions and people's knowledge about these practices, influences the public acceptance but in general the opinion is positive about water reuse for toilet flushing, clothes washing and garden irrigation, people are aware of the positive effects on the environment and the contribution to save scarce potable water sources.

However, careful studies and analysis of the consequences to the infrastructures of wide spread reuse systems must be done to avoid problems in the future because the reduction of usage increases the concentration of pollutant in the wastewater streams, which can have an impact on wastewater treatment. The main obstacles to a wide-spread implementation of these systems in buildings are, in one hand, the investment costs which are very high due to the treatment unit (except for simple direct reuse systems) and on the other hand the lack of a correct normalization and legislation to regulate this activity and guarantee a safe reuse. While practical guidelines for the reuse of greywater for irrigation are being made by government agencies involved in water management and regulation, there are a number of issues related to human, soil and plant health risks and environmental pollution that need further investigation to promote a safe reuse. There is an urgent need to examine these issues and concerns and seek answers to specific questions and develop guidelines with local data to ensure the sustainability of greywater reuse (Pinto et al., 2010).

So, the key challenges for these systems, in the future, are to make more affordable treatment units (which will probably happen once there will be more manufactures, if this practice becomes more common), and a more active role of governments in two fundamental areas: first, encouraging the implementation of these systems by creating incentives, like tax reductions and sponsorship for installation costs; second, creating legislation to guarantee a correct design and a safe reuse of treated water. There is also need to increase people's opinion about reusing greywater, which can be done by promoting educational campaigns to provide user-friendly information for people to know the best practices to avoid risks to human health, plant and soils.

REFERENCES

Alhumoud, J.M., Behbehani, H.S., Abdullah, T.H., (2003). Wastewater reuse practices in Kuwait. *The Environmentalist,* 23 (2), 117-126.

Al-Jayyousi O.R. (2003). Greywater reuse: towards sustainable water management. *Desalination* 156 (1-3):181-192.

Almeida, M. C., Baptista, J. M., Vieira, P., Moura e Silva, A.,Ribeiro, R. (2001). *O uso eficiente da água em Portugal no sector urbano: que medidas e que estratégias de implementação?* Jornadas de Engenharia Civil, Universidade do Minho. 09 -11 de Outubro, Guimarães, Portugal: 11 pp (in portuguese).

Almeida, M.C., Butler, D. and Friedler, E. (1999). At-source domestic wastewater quality. *Urban Water,* 1, 49–55.

Angelakis A.N., M, Marecos do Monte M.H.F., Bontoux L. and Asano T. (1999). The status of wastewater reuse practice in the Mediterranean basin: need for guidelines. *Water Research,* Vol. 33, No. 10, pp. 2201-2217.

AWWA-APHE-WPCF. 1989. *Standard methods for the examination of water and wastewater.* 17th ed.

Baggett, S., Jeffrey, P., Jefferson, B., (2006). Risk perception in participatory planning for water reuse. *Desalination* ,187 (1-3), 149-158.

Baumann, D., Kasperson, R., (1974). Public acceptance of renovated waste water: myth and reality. *Water Resources Research,* 10 (4), 667e673.

Bruvold, W.H., (1972). *Public Attitudes towards Reuse of Reclaimed Water,* vol. 137. Water Resource Centre, University of California, Berkeley, CA.

Bruvold, W.H., (1984). Obtaining public support for innovative reuse projects. In Proceedings of the *Water Reuse Symposium*, vol. 3, San Diego Ca, pp. 122–132.

Bruvold, W.H., (1988). Public opinion on water reuse options. *J. Water Pollution Control Federation*, 60(1) 45–50.

BSI (2010). BS 8525-1:2010 *Greywater systems - Part 1: Code of practice. British Standards Institution*. UK.

Burrows, W.D. Schmidt, M.O. Carnevale, R.M., and Shaub, S.A. (1991). Nonpotable reuse: development of health criteria and technologies for shower water recycle. *Water Science and Technology*, 24(9), 81–88.

Butler D., Friedler, E. and Gatt, K. (1995). Characterizing the quantity and quality of domestic wastewater. *Water Science and Technology*, 31:13–24.

Butler, D. (1991). A small-scale study of wastewater discharges from domestic appliances. *Water and Environment Journal*, 5: 178–184.

Casanova, L.M., Gerba, C.P. and Karpiscak, M. (2001). Chemical and microbial characterization of household greywater. *Journal Environmental Health Tox Hazardous Subst. Environ. Eng*., 36, (4), 395–401.

Chilton, J.C., Maidment, G.G., Marriott, D., Francis, A. and Tobias, G. (2000). Case study of a rainwater recovery system in a commercial building with a large roof. *Urban Water, 1*(4), 345-354.

Christova-Boal, D., Eden R.E. and McFarlane, S. (1996). An investigation into greywater reuse for urban residential properties. *Desalination*, 106, 391–397.

Clesceri L.S., Greenberg A.E. and Trussel R.R., (1989) *Standard Methods for the Examination of Water and Wastewater*, 17th Ed., American Public Health Association (APHA), American Water Works Association (AWWA), Water Pollution Control Federation (WPCF), Washington, DC, pp. 9-52 to 9-53.

Cogan, T.A., Bloomfield, S.F. and Humphrey, T.J. (1999).The effectiveness of hygiene procedures for prevention of cross contamination from chicken carcasses in the domestic kitchen. Letters in *Applied Microbiology*, 29, pp: 354-358.

Crook, J. (2003). An overview of water reuse. Keynote lecture at *International Seminar on Wastewater Reclamation and Reuse*. Organized by MED-REUNET (Mediterranean Network of Water Reclamation and Reuse). Sponsored by EU Research Directorates General, Agbar Foundation (Spain) and Ege University, Izmir, Turkey (http://www.med-reunet.com/02medr1/03_seminar.asp).

Crook, J., Okun, D., Pincince, A. (1994). *Water Reuse*, Report to WERF (Water Environment Research Foundation), Project 92-WRE-1, Alexandria, VA.

Denlay, J., Dowsett, B. (1994). Water Reuse the most reliable water supply available, Report prepared as part of the *Sydney Water Project*, Friends of the Earth Inc, Sydney, Australia, 57p.

Dishman, C.M., Sherrard, J.H., Rebhun, M. (1989). Gaining public support for direct potable water reuse. *Journal of Professional Issues in Engineering*, 115 (2), 154-161.

Dixon, A.M., Butler, D. and Fewkes, A. (1999). Guidelines for greywater reuse: health issues. *J. CIWEM*, 13: 322-326.

Dolnicar, S. and Schäfer, A.I. (2009). Desalinated versus recycled water: public perceptions and profiles of the accepters. *Journal of Environmental Management,* 90 (2), 888-900.

Domènech and Saurí (2010). Socio-technical transitions in water scarcity contexts: Public acceptance of greywater reuse technologies in the Metropolitan Area of Barcelona. *Resources, Conservation and Recycling*, 55 (2010) 53–62.

EA (2011). Greywater for domestic users: an information guide. *Environment Agency.* Bristol, UK.

EEA (2009). *Water resources across Europe - confronting water scarcity and drought.* EEA Report No 2/2009, 55pp. Copenhagen. ISBN 978-92-9167-989-8.

EEA (2012). *Towards efficient use of water resources in Europe.* European Environment Agency, EEA Report No1/2012, 68pp. Copenhagen. ISBN 978-92-9213-275-0.

EPA (1992). US Environmental Protection Agency, *Manual: Guidelines for Water Reuse*, Office of Water and Office of Research and Development, EPA/625/R-92/004, 1992.

Eriksson, E. and Donner, E. (2009). Metals in greywater: sources, presence and removal efficiencies. *Desalination*, 248 (1-3), pp: 271-278.

Eriksson, E., Auffarth, K., Henze, M. and Ledin, A. (2002). Characteristics of grey wastewater. *Urban Water*, 4, 85–104.

Flack, J.E., Greenberg, J., (1987). Public attitudes towards water conservation. *Journal of the American Water Works Association* 79 (3), 46e51.

Friedler, E. (2004). Quality of individual domestic greywater streams and its implication on on-site treatment and reuse possibilities. *Environmental Technology*, 25:997–1008.

Friedler, E. and Alfiya, Y. (2010). Physicochemical treatment of office and public buildings greywater. *Water Science and Technology, 62*(10), 2357-2363.

Friedler, E. and Butler, D. (1996). Quantifying the inherent uncertainty in the quantity and quality of domestic wastewater. *Water Science and Technology*, Volume 33, Issue 2, Pages 65-75,77-78.

Friedler, E. and Hadari, M. (2006). Economic feasibility of on-site greywater reuse in multi-storey buildings. Desalination, 190 (1-3), 221-234.

Friedler, E., Butler, D., Alfiya, Y. (2013), Wastewater composition. Chapter 17 of Source Separation and decentralization for Wastewater Management Tove A. Larsen, Kai M. Udert and Judit Lienert. Publication date: 01 Feb 2013 ISBN: 9781843393481 Pages: 520

Friedler, E., Kovalio, R. and Galil, N.I. (2005). On site greywater treatment and reuse in multi-storey buildings. *Water Science and Technology*, 51 (10): 187–194.

Friedler, E., Lahav, O., Jizhaki, H., Lahav, T., (2006). Study of urban population attitudes towards various wastewater reuse options: Israel as a case study. *Journal of Environmental Management* - Elsevier www.elsevier.com/locate/jenvman, 2005.

Friedler, E., R. Kovalio, and A. Ben-Zvi. (2006a). Comparative study of the microbial quality of greywater treated by three on-site treatment systems. *Environmental Technology* 27 (6):653-663.

Ghisi, E. and de Oliveira, S. M. (2007). Potential for potable water savings by combining the use of rainwater and greywater in houses in southern Brazil. *Building and Environment, 42*(4), 1731-1742.

Ghisi, E. and Ferreira, D. F. (2007). Potential for potable water savings by using rainwater and greywater in a multi-storey residential building in southern Brazil. *Building and Environment, 42*(7), 2512-2522.

Gilboa, Y. and Friedler, E. (2008). UV Disinfection of RBC- treated light greywater effluent: kinetics survival and regrowth of selected microorganisms. *Water Research*, 42, 1043–1050.

Godfrey, S., Labhasetwar, P. and Wate, S. (2009). Greywater reuse in residential schools in Madhya Pradesh, India – A case study of cost-benefit analysis. *Resources, Conservation and Recycling*, 53 (5), 287–293.

Gray, S.R. and Becker, N.S.C. (2002). Contaminant flows in urban residential water systems. *Urban Water*, 4 (4), pp: 331-346.

Gross, A., Kaplan, D. and Baker, K. (2007). Removal of chemical and microbiological contaminants from domestic greywater using a recycled vertical flow bioreactor (RVFB). *Ecological Engineering*, 31, pp: 107-114.

Gual, M., Moià A., and March, J.G. (2008). Monitoring of an indoor pilot plant for osmosis rejection and greywater reuse to flush toilets in a hotel. *Desalination* 219 (1-3):81-88.

Günther, F. (2000). Wastewater treatment by greywater separation: Outline for a biologically based greywater purification plant in Sweden. *Ecological Engineering*, 15, pp: 139-146.

Hartley, T.W. (2006). Public perception and participation in water reuse. *Desalination,* 187 115–126.

Hurlimann, A., (2007a). Attitudes to future use of recycled water in a Bendigo office building. *Water Journal of the Australian Water Association* 34 (6), 58-64.

Hurlimann, A. (2007b). Is recycled water use risky? An urban Australian community's perspective. *The Environmentalist* 27 (1), 83-94.

Hurlimann, A. (2007c). Recycling water for Australia's future e the case of two Victorian cities. In: *State of Australian Cities Conference*, 27,30 November, Adelaide, South Australia.

Hurlimann, A. (2008). Community Attitudes to Recycled Water Use: An Urban Australian Case Study e Part 2. Research Report No. 56. *Cooperative Research Centre for Water Quality and Treatment*, Adelaide, SA.

Hurlimann, A. and McKay, J., (2004). Attitudes to reclaimed water for domestic use: Part 2. Trust. Water. *Journal of the Australian Water Association,* 31 (5), 40-45.

Hurlimann, A., McKay, J. (2007). Urban Australians using recycled water for domestic non-potable use—An evaluation of the attributes price, saltiness, colour and odour using conjoint analysis. *Journal of Environmental Management,* 83 (2007) 93–104.

Hurlimann, A.C., Hemphill, E., McKay, J., and Geursen G. (2008). Establishing components of community satisfaction with recycled water use through a structural equation model. *Journal of Environmental Management* 88 (4), 1221-1232.

Imteaz, M.A., Shanableh, A., Rahman, A. and Ahsan, A. (2011). Optimisation of rainwater tank design from large roofs: A case study in Melbourne, Australia. *Resources, Conservation and Recycling, 55*, 1022-1029.

IPQ. (2005). NP 4434 - *Reutilização de águas residuais urbanas tratadas na rega*. Instituto Português da Qualidade (in portuguese). Lisboa, Portugal.

Jefferson, B., Palmer, A., Jeffrey, P., Stuetz, R. and Judd. S. (2004). Grey water characterization and its impacts on the selection and operation of technologies for urban reuse. *Water Science and Technology*, 50 (2), 157–164.

Jeffrey, P., Jefferson, B. (2003). Public receptivity regarding in-house water recycling: results from a UK survey. *Water Science and Technology— Water Supply* 3 (3), 109–116.

Jeffrey, P., Temple, C. (1999). Sustainable water management: some technological and social dimensions of water recycling. *Sustainable Development International* 1, 63–66.

Kantanoleon N., Zampetakis L., Manios T. (2007). Public perspective towards wastewater reuse in a medium size, seaside, Mediterranean city: A pilot survey. *Resources, Conservation and Recycling,* 50: 282–292.

Khan S.J., Gerrard L.E., (2006). Stakeholder communications for successful water reuse operations. *Desalination,* 187: 191-202.

Laak, (1974). Relative pollution strengths of undiluted waste materials discharged in households and the dilution waters used for each. *Manual of grey water treatment practice* (pp. 68-78). Michigan, USA: AnnArbor.

Lohman, L.C., Milliken, J.G. (1985). *Informational/educational Approaches to Public Attitudes on Potable Reuse of Wastewater.* U.S. Department of the Interior, Denver.

Macpherson and Law (2003), Winning minds to water, reuse: The road to NEWater. In: *19th Annual WateReuse Symposium*: Water Reuse — A Resource Without Boarders, WateReuse Association, San Antonio, Texas, USA.

Mandal, D., Labhasetwar, P., Dhone, S., Dubey, A. S., Shinde, G. and Wate S. (2011). Water conservation due to greywater treatment and reuse in urban setting with specific context to developing countries. *Resources Conservation and Recycling, 55* (3), 356-361.

Manville, D. ,Kleintop, E. , Miller, B., Davis, E. , Mathewson, J. and Downs, T. (2001). Significance of indicator bacteria in a regionalized wastewater treatment plant and receiving waters. *International Journal of Environment and Pollution,* 15 (4), pp: 461-466.

March, J.G., Gual M., and Orozco F. (2004). Experiences on greywater re-use for toilet flushing in a hotel (Mallorca Island, Spain). *Desalination,* 164 (3):241-247.

Marecos do Monte, M. H. (1996). *Contributo para a Utilização de Águas Residuais tratadas para Irrigação em Portugal* (in Portuguese). LNEC, Lisbon.

Marks, J., (2004). Back to the future: reviewing the findings on acceptance of reclaimed water. Conference Proceedings, *Enviro04,* Australian Water Association, 28–31 March 2004, Sydney, Australia.

Marks, J., Cromar, N., Fallowfield, H., Oemcke, D. (2003). Community experience and perceptions of water reuse. *Water Science and Technology-Water Supply,* 3 (3), 9–16.

Marks, J.S. (2006). Taking the public seriously: the case of potable and non-potable reuse. *Desalination* 187: 137–147.

Marks, J.S., Martin, B., Zadoroznyj, M. (2006). Acceptance of water recycling in Australia: national baseline data. *Water,* 33 (2), 151-157.

Matos, C. (2009). Reutilização de água: Utilização de águas cinzentas insitu. PhD Thesis. Universidade de Trás-os-Montes e Alto Douro. Pp: 167.

Matos, C., Friedler, E., Monteiro, A., Rodrigues, A., Teixeira, R., Bentes, I. and Varajão, J. (2013). Academics perception towards various water reuse options: University of Trás-os-Montes e Alto-Douro - UTAD Campus (Portugal) as a case study. *Urban Water.* DOI: 10.1080/1573062X.2013.775314.

Matos, C., Sampaio, A. and Bentes, I. (2010). Possibilities of greywater reuse in non-potable in situ urban applications, according with its quality and quantity. *WSEAS Transactions on Environment and Development* 7 (6), pp: 499-508.

Matos, C., Sampaio, A:, Bentes, I. (2012). "Irrigation - Water Management, Pollution and Alternative Strategies", book edited by Iker García-Garizábal and Raphael Abrahao, ISBN 978-953-51-0421-6, Published: March 28, 2012 under CC BY 3.0 license- Chapter

9: Greywater Use in Irrigation: Characteristics, Advantages and Concerns. http://www.intechopen.com/books/irrigation.

McKay, J., Hurlimann, A.C. (2003). Attitudes to reclaimed water for domestic use: Part 1. Age. Water. *Journal of the Australian Water Association,* 30 (5), 45-49.

Menegaki, A., Hanley, N., Tsagarakis, K.P. (2006). Social acceptability and evaluation of recycled water in Crete: a study of consumers' and farmers' attitudes. *Ecological Economics.* 62 (1), 7e18.

Metcalf and Eddy., George Tchobanoglous, Franklin L. Burton, and H. David Stensel (2003). *Wastewater engineering: treatment and reuse*. 4th ed, *McGraw-Hill series in civil and environmental engineering*. Boston: McGraw-Hill.

Muthukumaran, S., Baskaran, K., Sexton, N. (2011). Quantification of potable water savings by residential water conservation and reuse – A case study. *Resources, Conservation and Recycling, 55*(11), 945-952.

Nancarrow, B., Leviston, Z., Po, M., Porter, N., Tucker, D. (2008). What drives communitie's decisions and behaviours in the reuse of wastewater. *Water Science and Technology.* 57 (4), 485-491.

Nolde, E. (1999). Greywater reuse for toilet flushing in multi-storey buildings – Over ten years experience in Berlin. *Urban Water* 1: 275-284.

Olson, B.H., Henning, J.A., Marshack, R.A., Rigby, M.G. (1979). Educational and social factors affecting public acceptance of reclaimed water. In: *Water Reuse Symposium*, Denver, Colorado, pp. 1219-1231.

Ottoson, J. and Stenström, T.A. (2003). Faecal contamination of greywater and associated microbial risks. *Water Research*, 37, pp: 645–655.

Paris, S. and Schlapp, C. (2010). Greywater recycling in Vietnam – Application of the HUBER MBR process. *Desalination*, 250 (3), 1027-1030.

Pinto, U., Maheshwari, B.L. and Grewal, H.S. (2010). Effects of greywater irrigation on plant growth, water use and soil properties. *Resources Conservation and Recycling, 54* (7), 429-435.

PNUEA (2001). *Programa Nacional para o Uso Eficiente da Água* (in portuguese). MAOT-IA Lisbon.

Po, M., Nancarrow, B.E., Leviston, Z., Poter, N.B., Syme, G.J., Kaercher, J.D. (2005). Predicting Community Behaviour in Relation to Wastewater Reuse: What Drives Decisions to Accept or Reject? *Water for a Healthy Country National Research Flagship.* CSIRO Land and Water, Perth, WA.

Robinson, K.G., Robinson, C.H., Hawkins, S.A. (2005). Assessment of public perception regarding wastewater reuse. *Water Science and Technology: Water Supply*, 5 (1), 59-65.

Rose, J.B., Gwo-Shing, Sun., Gerba, C.P., and Sinclair, N.A. (1991). Microbial quality and persistence of enteric pathogens in greywater from various households sources. *Water Research*, 25 (1), 37–42.

Santos, C. (2012). *Otimização ambiental do uso de água em edifícios* (in portuguese). PhD Thesis. Faculdade de Engenharia, Universidade do Porto.

Santos, C. and Taveira-Pinto, F. (2013). Analysis of different criteria to size rainwater storage tanks using detailed methods. *Resources, Conservation and Recycling, 71*, 1-6.

Santos, C., Taveira-Pinto, F., Cheng, C.Y. and Leite D. (2011). Development of an experimental system for greywater reuse. *Desalination*, 285, 301-305.

Shin, H.S., Lee, S-M., Seo, I.S., Kim, G.O., Lim, K.-H. and Song, J.S. (1998). Pilot scale SBR and MF operation for the removal of organic and nitrogen compounds from greywater. *Water Science Technology*, 38 (6), pp. 79-88.

Siegrist, R., Boyle, W.C. and Witt, M. (1976). Characteristics of rural household wastewater. *Journal of the Environmental Engineering Division*, 102 (3), 533–548.

Simpson, J.M. (1999). Changing community attitudes to potable reuse in South-East Queensland. *Water Science and Technology*, 40 (4–5), 59–66.

Standards Australia (2001). HB 326-2008 *Urban Greywater Installation Handbook for Single Households*. NSW, Australia.

Stone, R., Kahle, R. (1974). *Wastewater Reclamation. Socio Economics, Technology and Public Acceptance*. Office of Water Resource Research, US Department of the Interior, Washington, DC.

Surendan, S. and Wheatley, D.A. (1998). Greywater reclamation for non-potable reuse. *J. CIWEM*, 12: 406–413.

Tchobanoglous, G., Burton, F.L. and Stensel, H.D. (2003). *Wastewater Engineering: treatment and reuse*. New York: Mcraw-Hill, 1819 pp.

Travis, M.J., Weisbrod, N. and Gross, A. (2008). Accumulation of oil and grease in soils irrigated with greywater and their potencial role in soil water repellency. *Science of the Total Environment*, 394 (1), pp: 68-74.

Tsagarakis, K.P., Mellon, R., Stamataki, E., Kounalaki, E. (2007). Identification of recycled water with an empirically derived symbol increases its probability of use. Environmental Science and Technology 41 (20), 6901e-6908.

Van der Hoek, J.P., Dijkman, B.J., Terpstra, G.J., Uitzinger, M.J., van Dillen, M.R.B. (1999). Selection and evaluation of a new concept of water supply for "Ijburg Amsterdam. *Water Science and Technology* 39 (5), 33-40.

Wiel-Shafran, A., Ronen, Z., Weisbrod, N., Adar, E. and Gross, A. (2006). Potential changes in soil properties following irrigation with surfactant-rich greywater. *Ecological Engineering*, 26 (4), pp: 348-354.

In: Water Conservation
Editor: Monzur A. Imteaz

ISBN: 978-1-62808-993-6

Chapter 7

REUSE OF LAUNDRY GREYWATER IN IRRIGATION: POTENTIAL CHANGES IN SOIL PARAMETERS

***A. H. M. Faisal Anwar*[1,*] *and Monzur Alam Imteaz*[2]**
[1]Department of Civil Engineering, Curtin University, Perth, Australia
[2]Faculty of Engineering and Industrial Sciences, Swinburne University of Technology, Melbourne, Australia

ABSTRACT

Laundry greywater is potentially surfactant-rich and thus have significant impact on soil hydrologic properties. This study investigates the potential changes of soil hydrologic parameters following irrigation of laundry greywater in two types of soils. The soil parameters investigated include hydraulic conductivity, pH, electrical conductivity (EC), porosity, bulk density and capillary pressure-saturation relationship. A soil column composed of several PVC rings was used for greywater irrigation. The experiments were conducted under unsaturated condition for different synthetic greywater concentration. In each experiment, pH, EC, hydraulic conductivity, soil porosity and bulk density were measured. At the end of each experiment, the column was dismantled and moisture content in each ring was measured gravimetrically and the soil suction in each ring was taken as the distance from the reference groundwater table. The breakthrough curves with EC and pH measured at the column outlet revealed that the greywater with higher concentration reach the column outlet faster because of increased hydraulic conductivity. The soil hydraulic conductivity was found steadily increasing with greywater concentration for both soils. Pressure saturation curves showed that the capillary rise decreases with greywater concentration due to the reduction of surface tension. However, the residual water content in soil B was found higher than soil A following greywater irrigation.

Keywords: Greywater, soil, irrigation, surfactant, hydraulic conductivity

---

* Corresponding author: A. H. M. Faisal Anwar. E-mail: f.anwar@curtin.edu.au.

INTRODUCTION

As forecasted, arid and semi-arid regions may face enormous shortage of potable water in the coming decades because of the global climate change. People are more becoming aware to preserve water and greywater reuse in lawn/gardening and parkside irrigation may be one of the best options to save potable water.

Greywater is a non-toilet component of household wastewater generated from bathtubs, showers, sinks, washing machines and dishwashers. Among these sources, laundry greywater may be considered as one of the best choices for reuse in irrigation because greywater originating from other sources (especially kitchen) may contain oil and grease (Travis et al., 2008 and Friedler, 2004).

Recycling of greywater in irrigation is becoming common in many countries including Australia. In a typical household in Western Australia, total generation of greywater is 117 l/d per person (DHWA, 2005). Laundry and bathroom greywater contributes about 89% of this volume, which can be reused for the lawn/gardening watering. The domestic wastewater may contain reduce level of nitrogen, solids and organic matter, but often it contains higher level of surfactants, oils, boron and salt (Gross et al., 2005; Shafran et al., 2005 and Shafran et al., 2006). These components of greywater may have harmful effects on soil, plants and underground water. Surfactants are the major components of domestic detergents which have hydrophilic head and hydrophobic tail (Shafran et al., 2005; Shafran et al., 2006 and Abu-Zreig et al., 2003). The hydrophobic group contains a long alkyl chain of C_{10}–C_{20}. The hydrophilic group has an electrical charge, or is polarized, and can form hydrogen bonds as shown in Figure 1 (Shafran et al., 2005).

One of the main physical characteristics of surfactant (surface active agent) molecules in aqueous solution is that it accumulates onto different interfaces (such as, liquid/liquid; liquid/air or solid/liquid interfaces). This increases the distance between the water molecules and therefore causes the reduction of surface tension (Anwar et al., 2000 and Anwar, 2011). The capillary rise (h) is directly related to the surface tension as described by the Young–Laplace equation (Equation 1).

$$h = \frac{2\sigma \cos \alpha}{\rho g r} \tag{1}$$

where σ is the surface tension of the imbibing solution in N/m, α is the contact angle in degrees, ρ is the density of the liquid in kg/m^3, g is the gravity force in m/s^2 and r is the capillary radius in m.

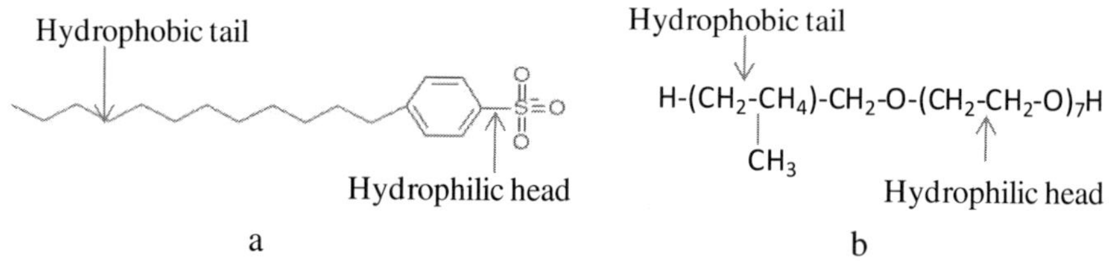

Figure 1. Demonstration of surfactant components (a) linear alkyl benzene sulphonates (anionic surfactant), (b) linear primary Alcohol Ethoxylate (AE) (nonionic surfactant).

The surfactant molecules present in surfactant-rich greywater accumulates onto the air-water interface which causes the reduction of surface tension of aqueous solution. The reduction of surface tension in surfactant-rich greywater may change the underlying soil structure and thus the soil-water environments, which are ignored in most of the cases. Interaction between laundry greywater and the saturated soil has been studied to quantify the soil hydraulic conductivity (Misra and Sivongxay, 2009) in different soil samples. Shafaran et al. (2005) suggested that accumulation of surfactants from greywater may lead to water repellent soil with significant impact on agricultural productivity and environmental sustainability.

Travis et al. (2010) conducted a controlled study in containers using three types of soils planted with lettuce which were irrigated with freshwater, raw artificial greywater and treated artificial greywater. They found that the soil irrigated with raw greywater shows more water repellency than the soil irrigated with freshwater or treated greywater. Recently Anwar (2011 and 2012a) investigated the effects of greywater irrigation on soil properties for one type of soil and Anwar (2012b) studied the effect of greywater irrigation on air-water interfacial area. This chapter presents the potential changes of soil hydrologic parameters for two types of soil following greywater irrigation. The experiments are conducted in a special type of soil column consists of different rings representing actual subsurface aquifer system (Anwar et al., 2000 and Anwar, 2011).

Material and Methods

Materials

Two types of soil media was chosen for this study. The physical properties of the porous medium are shown in Table 1. The soil media were washed for several times using tap water and oven dried (at 105°C for 24 hrs) before sieve analyses (Method AS1289). The same sieve diameters were used in sieve analysis for both soils.

The DUO 2X Ultra Concentrated Top Loader Aromatic Detergent Powder is selected as laundry detergent for greywater preparation based on the availability of required information in the literatures (Lanfax Lab, 2012). The concentrations of laundry greywater were calculated based on the full washing capacity of a washing machine and following the given instructions of DUO washing powder. Normal load wash and three cycles of wash and rinse were assumed and the greywater concentrations used in the experiments were, 0.26 g/L, 0.316g/L, 0.368g/L, 0.442g/L. 0.6g/L and 0.736g/L. The pH and EC of these greywater and tap water is shown in Table 2.

Table 1. Physical properties of soil used in this study

Soil	Particle range (mm)	Median grain d_{50} (mm)	Porosity, n	Bulk density, ρ_b (g/cm^3)	Hydraulic conductivity Ks (m/day)
A	0.075-2.36	0.38	0.4	1.68	38.48
B	0.075-2.36	0.45	0.39	1.58	32.306

Table 2. The pH and EC value of different greywater and tap water

Greywater concentration (g/L)	Tap Water	0.26	0.316	0.368	0.442	0.6	0.736
pH	7.36	9.78	10.11	10.18	10.26	10.32	10.39
EC (μs/cm)	468	885	959	976	1039	1430	1639

Column Experiments

The experiments were conducted in a laboratory soil column composed of several PVC rings of 9 cm inner diameter and 3 cm length as shown in Figure 1. The column was packed successively with the selected soil medium in small increments under water saturated condition and tapped at the bottom. This procedure ensured the elimination of any trapped air and layer formation during the packing process. The effective length of the soil column was 54cm. The column was kept saturated for 1 hour and the outlet tank was brought down and kept the water level of the outlet tank at the same level of the bottom of the soil column. This arrangement could maintain the water level at the bottom of the soil column referring to groundwater table. The column was kept in this position for 24 hours to equilibrate the system. After establishing equilibrium in the system, the greywater of known concentration was flushed through the column and pH and EC at column outlet were measured in every two minutes interval. Total pore volume of the column was calculated as 1443cm^3. Greywater irrigation continued until the EC of the column effluent became constant. To get constant EC at the column outlet, approximately 7 number of pore volume of greywater solution was flushed into the column for 120 minutes. The flow velocity of greywater flushing was on an average 3.11 cm/min. The column was again kept 24 hrs to establish equilibrium in the system and then the column was dismantled (Anwar et al., 2000). The moisture content in each ring was measured gravimetrically and the suction head corresponding to each ring was taken as the distance between the ring's mid-point and the bottom water level (Anwar et al., 2000). The same experiment was repeated for different greywater concentration and one experiment was done with tap water at the beginning to record the initial soil properties for both soil types. The porosity of soil was taken as the saturated moisture content and the bulk density was calculated for each ring gravimetrically. Separate experiments were done with the same greywater concentration to calculate the soil hydraulic conductivity using constant head method. All the experiments were conducted at 22±1^{0}C.

RESULTS AND DISCUSSION

Effect on Electrical Conductivity

Higher concentration of greywater contributes to higher Electrical Conductivity (EC) as given in Table 2. The EC of aqueous solution indicates the presence of salt and hence the salinity of the soil. The EC of tap water was found in this study as 468 μs/cm, but sometimes it may go up to 500 μs/cm plus.

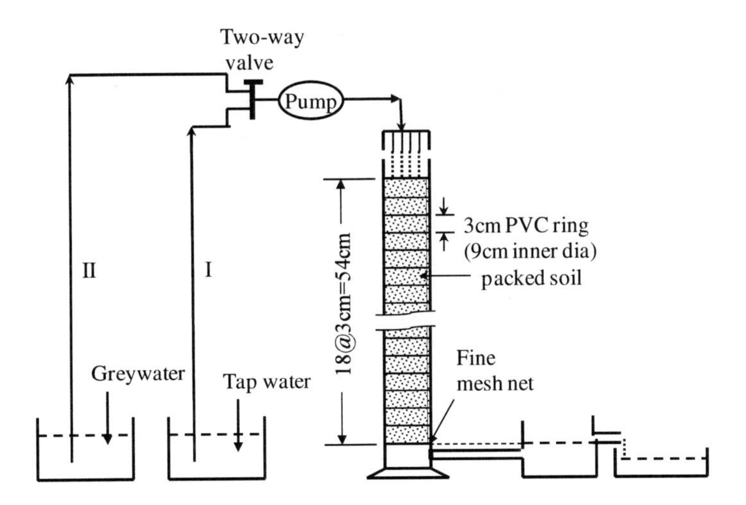

Figure 1. Experimental setup.

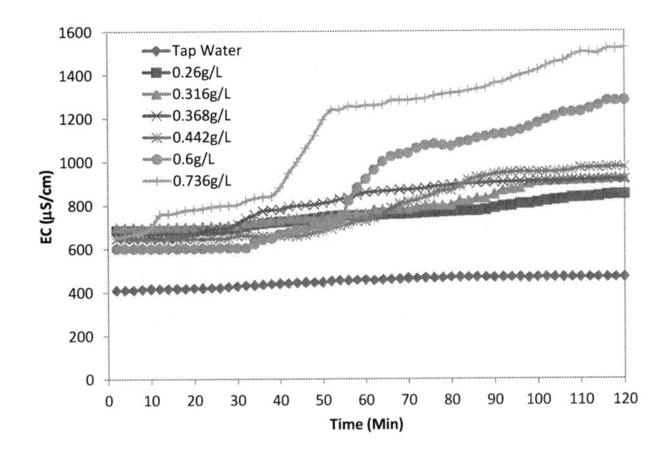

Figure 2. Column effluent electrical conductivity of soil A.

The salinity is an issue where laundry water is discharged over the soil in gardens and lawns. A salt is simply a compound that dissociates in water to form cations and anions. Salt in the washing powder is made up of cations such as, calcium, magnesium, and potassium and anions of sulphate, phosphate, nitrate, chloride and carbonate (Shafran et al., 2005). The breakthrough curves (BTC) of EC measurements for soil A and B are shown in Figure 2 and Figure 3 respectively.

The BTCs indicate that the higher concentration greywater reach the column outlet faster but it took longer time in soil B than soil A. This is because of the hydraulic conductivity of soil B which was less than soil A and thus the solute took longer time in soil B to reach the column outlet.

The normal electrical conductivity level for irrigation water is about 1ds/m while higher EC values are more likely to induce loss of plant production. The general plant health will be affected by increased salinity in irrigation water due to effects of increased salinity on the physiology of the plant and the effects of soil salts (Anwar, 2012a).

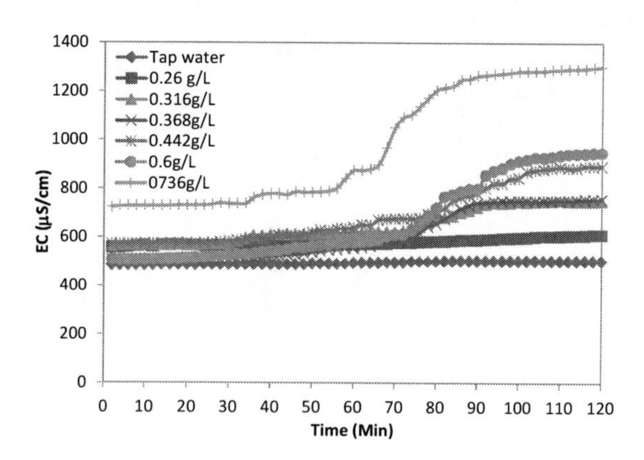

Figure 3. Column effluent electrical conductivity of soil B.

The normal salinity level in laundry greywater is found between 800 µs/cm to 1050 µs/cm for normal load wash (one scoop of washing powder) and rinse water together. Most of cities discharge greywater into ocean without reuse it; therefore much attention was not paid on the effects of greywater on soil structures. Appropriate regulations are absent in most places to reuse laundry greywater. However, diluting laundry greywater before irrigation may be needed to maintain low levels of soil salinity.

Effect on pH

The pH values in column effluent were measured with time for each experiment and shown in Figure 4 and Figure 5 for soil A and B respectively. Tap water pH value was measured 7.36. The pH value for different greywater concentration in this study varies between 9.78-10.39.

Results revealed that the pH of effluent in soil A column starts to increase after 40 minutes for the lower range of greywater concentration but it was obtained much faster in the higher concentration. But interestingly, the pH of effluent in soil B column started to decrease slightly. It also revealed that the pH of initial effluent in soil B was higher than the effluent of soil B. This finding provides the indication that the soil B was more acidic than soil A. However, soil acidity was not tested for the selected soil in this study before column experiment. Higher greywater concentration with high pH value has the ability to dissolve organic matter, such as sweat, blood, food and also has adverse effect on skin (Lanfax Lab, 2012).

Therefore, one should carefully handle the wash water with pH > 10. The pH value is related with soil structure and health of the plants. High pH value could lead to the dissolution of organic material and induces dispersion in the soil. This is because the detergent in the washing powder supposes to remove soils from clothes. Therefore high pH liquids act as dispersing agents, causing the soil particles to separate and lead to soil structure decline. The normal pH range for biological activity is in between 5 to 9. If pH value reached more than 9, biological activity won't be happened in normal way and dissolved organic material can be leached out of soil (Lanfax Lab, 2012). The dissolved organic materials degrade in time to become plant nutrients, so a loss of organic material may be detrimental to plant health (Lanfax Lab, 2012). However, laundry greywater irrigation is more suitable for acidic soil.

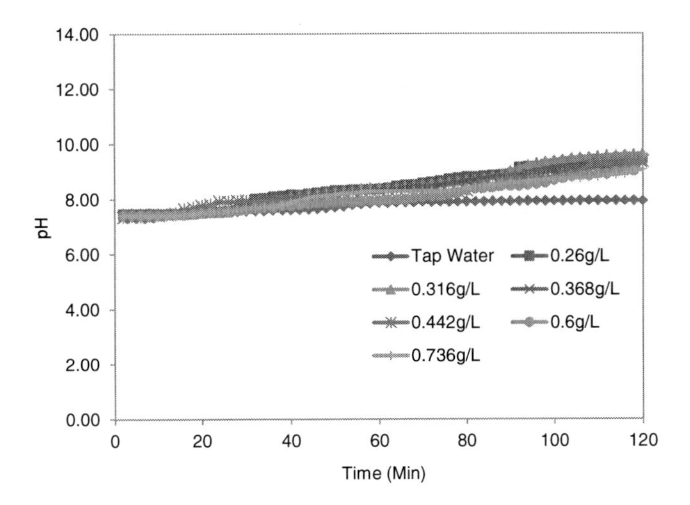

Figure 4. Changes of pH in soil A.

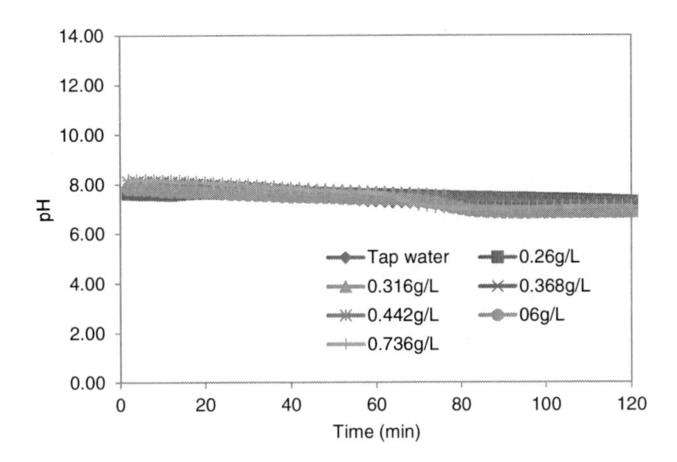

Figure 5. Changes of pH in soil B.

Hydraulic Conductivity

Soil hydraulic conductivity depends on the type of soil, porosity and configuration of the soil pores. Saturated hydraulic conductivity data for sequential leaching of soil with tap water and laundry greywater are shown in Figure 6 for soil A and B. Hydraulic conductivity increases steadily with greywater concentration. This is because of the washing of very fine particles and reduction of surface tension. Increasing hydraulic conductivity means overall water transmitting capacity increases and the water retention capacity decreases. Hydraulic conductivity is a very important soil property when determining the potential risks for widespread groundwater contamination by a contaminating source. Soil with high hydraulic conductivities and large pores for transmitting water are likely candidates for far reaching contamination. This means increasing hydraulic conductivity due to greywater irrigation may enhance groundwater contamination. However, the hydraulic conductivity of soil A was initially higher than B and it remains higher for different greywater irrigation.

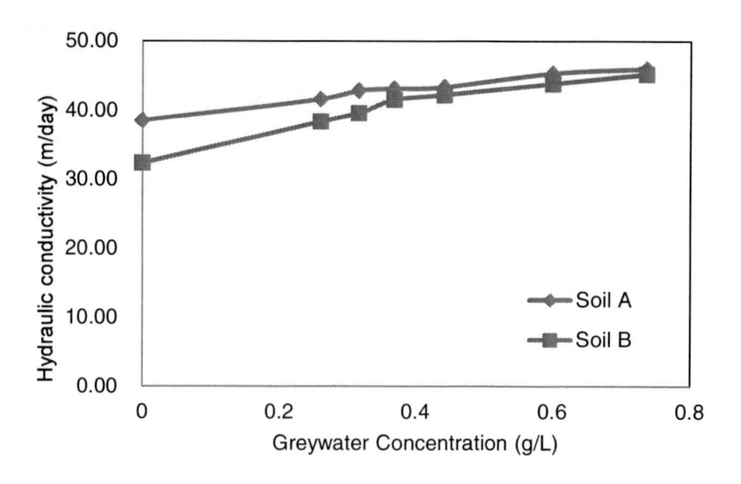

Figure 6. Changes of soil hydraulic conductivity following greywater irrigation.

Table 3. Soil bulk density and porosity

Soil type	Greywater concentration (g/L)	Tap Water	0.26	0.316	0.368	0.442	0.6	0.736
Soil A	Bulk Density ρ_b, (g/cm^3)	1.680	1.65	1.61	1.61	1.66	1.66	1.63
Soil B	Bulk Density ρ_b, (g/cm^3)	1.58	1.58	1.57	1.57	1.56	1.57	1.57
Soil A	Porosity, n (cm^3/cm^3)	0.40	0.40	0.42	0.43	0.41	0.41	0.40
Soil B	Porosity, n (cm^3/cm^3)	0.39	0.38	0.40	0.40	0.40	0.40	0.40

Soil Bulk Density and Porosity

The soil bulk density and porosity were measured after each experiment with greywater and tap water irrigation and shown in Table 3. Results revealed that the greywater irrigation does not have significant effect on these soil properties. Both of these soils have similar porosity but their hydraulic conductivities are different. Well graded soil provides lower hydraulic conductivity than the uniform soil. Though the soil grain ranges were similar but their grain size distributions were dissimilar. This indicates soil B was well graded than soil A.

Pressure-Saturation Relationship

After each column experiment for different greywater irrigation, the moisture contents in each ring were measured gravimetrically and the capillary pressure head was taken as the distance between the groundwater table and the mid-point of each ring (Anwar et al., 2000). The pressure-saturation curves for each experiment are plotted in Figure 7 for soil A and Figure 8 for soil B respectively.

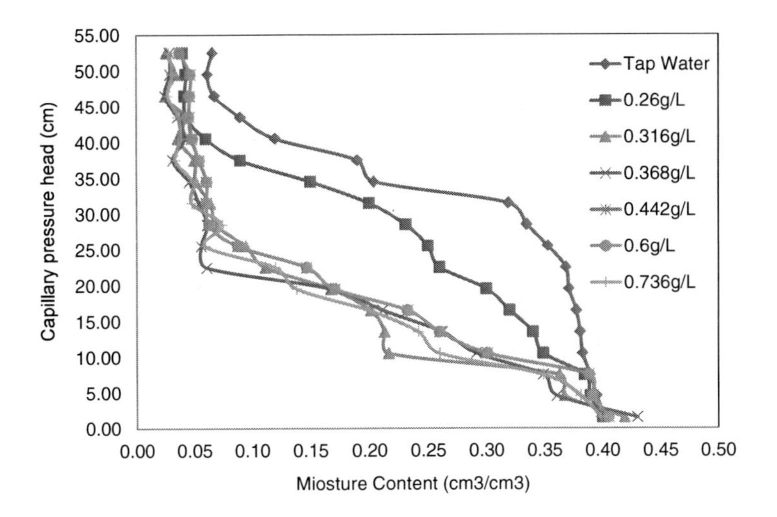

Figure 7. Pressure-saturation relationship for greywater irrigtaion in in soil A.

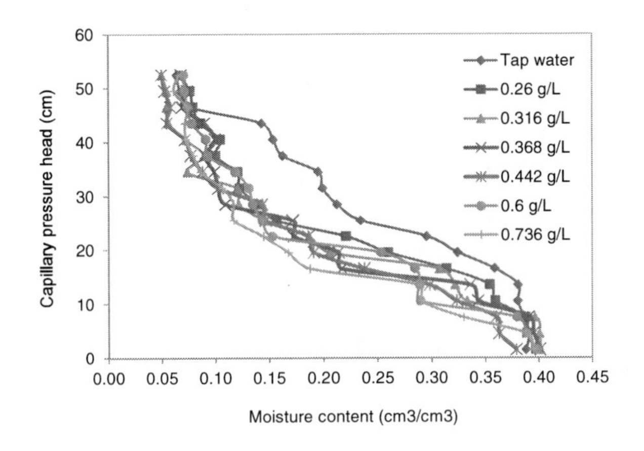

Figure 8. Pressure-saturation relationship for greywater irrigtaion in in soil B.

Results revealed that the residual moisture content in soil B becomes higher than soil A after greywater irrigation. This is because that the hydraulic conductivity is higher in soil A than B which indicates that the water retention capacity in soil B becomes higher following greywater irrigation. Capillary rise is a phenomenon that can have both beneficial and detrimental effects on the soil. It is a main mechanism by which plants can draw water from below the root zone, but it is also a mechanism contributing to the accumulation of salts in the soil (Shafran, et al., 2005). The reduction of capillary pressure with increasing greywater concentration indicates the presence of surfactants in greywater. Considering negligible contact angle, it is expected that the surfactant would reduce the surface tension of the aqueous phase. This is because of the surfactant monomers that have an affinity to accumulate onto the interfaces. This phenomenon consequently changes the migration pattern in the soil pores. Figure 7 and 8 revealed that the capillary pressure decrease sharply at the lower concentration of greywater but have a little effect at the higher concentration. This is because the surfactant monomers form micelles when the entire interfaces are saturated with monomers.

At this moment, the surfactant concentration is known as critical micelle concentration (CMC). As the specific surfactant in the selected detergent is unknown in this study, it is hard to find CMC of the surfactant present in the laundry detergents.

Though the surfactant in DUO detergent is unknown but several studies have shown that anionic and non-ionic surfactants are present in detergents (Shafran, et al., 2005; Misra and Sivongxay, 2009). Shafran, et al. (2005) reported that it is only the surfactant in laundry detergent has the influence on the reduction of surface tension, not any other ingredients present in it.

Another explanation for decrease in capillary pressure is embedded in the mechanism of surfactant adsorption onto the soil surfaces. Shafran, et al. (2005) performed experiments with the pure surfactants normally present in laundry detergent such as linear alkylbenzene sulfonate (LAS) and found that the electrostatic bonds of negatively charged sulfonate groups interact with the positively charged sand surfaces causing the adsorption of hydrophobic tails of LAS monomers and protruding into the aqueous phase.

These actions of surfactants enhance the soil to be water-repellent. A water-repellent soil (or hydrophobic soil) does not wet up spontaneously when a drop of water is placed upon the surface and thus it becomes a problem for future irrigation and enhanced environmental pollution. However, the current knowledge on the causes and characteristics of water-repellent soil is insufficient and further study should be undertaken to understand the accumulation of surfactant in soil due to greywater irrigation.

Conclusion

A series of soil column experiments were conducted to investigate the possible changes of soil parameters during the reuse of laundry greywater in irrigation in two types of soil. The column was constructed using several PVC rings and packed with soil under water-saturated condition. The column experiments were performed for different greywater concentrations under unsaturated condition.

During each experiment, pH and EC value was measured at the column outlet. At the end of each experiment, column was dismantled and the moisture content in each ring was measured gravimetrically and corresponding soil suction was measured. After each experiment, soil bulk density and porosity were also measured and found that greywater irrigation does not have significant effect on these parameters. Hydraulic conductivity was measured for each greywater concentration for both soils using constant head method and found that it increases with greywater concentration.

The results of EC and pH revealed that the greywater reaches faster at the column outlet with increasing concentration because of the steady increase of soil hydraulic conductivity. Capillary pressure –saturation relationship revealed that the capillary rise decreases with greywater concentration because of the surface tension reduction. Adsorption of surfactant onto soil surfaces may enhance the soil to be water-repellent but further study is needed for greywater adsorption onto soil.

REFERENCES

Abu-Zreig, M., Rudra, P. R., Dickinson, T. W. (2003). Effect of application of surfactants on hydraulic properties of soils, *Biosyst. Eng*., 84(3), 363–372.

Anwar, A. H. M. F. (2012a). Reuse of laundry greywater in irrigation and its effects on soil hydrologic parameters, In: *Proc. of Int. Conf. on Future Environment and Energy (ICFEE 2012*), Singapore, 26-28 February, 16-20.

Anwar, A. H. M. F. (2012b). Effect of greywater irrigation on air-water interfacial area in porous medium, *Int. J. of Civil and Environ. Eng*., 6, 35-39.

Anwar, A. H. M. F. (2011). Effect of laundry greywater irrigation on soil properties, *J. of Environ. Res. and Develop.,* 5(4), 863-870.

Anwar, A. H. M. F., Bettahar, M., Matsubayashi, U. (2000). A method for determining air-water interfacial area in variably saturated porous media. *J. of Contam. Hydrol*., 43(2), 129-146.

DHWA (Department of Health Western Australia), 2005. *Code of Practice for the reuse of greywater in Western Australia.*

Friedler, E., (2004). Quality of individual domestic greywater streams and its implication for on-site treatment and reuse possibilities, *Environ. Technol*. 25(9), 997–1008.

Gross, A., Azulai, N., Oron, G., Ronen, Z., Arnold, M., Nejidat, A. (2005). Environmental impact and health risks associated with greywater irrigation: a case study, *Water Sci. and Technol*., 52(8), 161–169.

Lanfax Lab, http://www.lanfaxlabs.com.au/ (accessed July 15, 2012)

Misra, R. K., Sivongxay, A., (2009). Reuse of laundry greywater as affected by its interaction with saturated soil, *J. of Hydrol*., 366(1-4), 55–61.

Shafran, A. W., Gross, A., Ronen, Z., Weisbrod, N., Adar, E., (2005). Effects of surfactants originating from reuse of greywater on capillary rise in the soil, *Water Sci. and Technol*., 52(10-11), 157-166.

Shafran, A. W., Ronen, Z., Weisbrod, N., Adar, E., Gross, A. (2006). Potential changes in soil properties following irrigation with surfactant-rich greywater, *Ecol. Eng*. 26(4) 348–354.

Travis, M. J., Shafran, A. W., Weisbrod, N., Adar, E., Gross, A. (2010). Greywater reuse for irrigation: Effect on soil properties, *Sci. Total Environ*., 408(12), 2501–2508.

Travis, M. J., Weisbrod, N., Gross, A. (2008). Accumulation of oil and grease in soils irrigated with greywater and their potential role in soil water repellency, *Sci. Total Environ*., 394 (1), 68-74.

In: Water Conservation
Editor: Monzur A. Imteaz

ISBN: 978-1-62808-993-6

Chapter 8

IMPACTS OF WATER PRICE AND RESTRICTIONS IN WATER DEMAND: A CASE STUDY FOR AUSTRALIA

Md. Mahmudul Haque,[1] Amir Ahmed[2] and Ataur Rahman[1,*]
[1]University of Western Sydney, Australia
[2]EnviroWater Sydney, Australia

ABSTRACT

Adequate water supply is a prerequisite to a nation's existence and economic growth. Water, although, available in many different forms, potable water is scarce in many countries. To ensure a sustainable water supply to general public, in addition to building water reservoirs/dams and/or maintaining a sustainable groundwater aquifer system, water pricing and demand management are also essential. Right water pricing can offer number of advantages such as cost recovery and long term demand reduction. This chapter examines how water price and demand restriction can assist in water conservation in potable urban water supply. It has been found that water usage price is nearly inelastic in Australia. On the contrary, water restrictions have played an important role in Australia as a drought response option in reducing water consumption during drought periods when reservoir water level becomes quite low. It has been found that water restriction can reduce water demand during the restriction period by about 33% in Australia. As these restrictions have some negative impacts on the community, this option needs to be carefully investigated in terms of cost and benefits against some other alternative means such as supply augmentation, water recycling and rainwater harvesting.

Keywords: Water price, water restriction, demand management, water recycling, rainwater

* Corresponding author: Ataur Rahman. School of Computing, Engineering and Mathematics, University of Western Sydney, Building XB, Kingswood, Locked Bag 1797, Penrith, NSW 2751, Australia. E-mail: a.rahman@uws.edu.au.

INTRODUCTION

Water is a vital source for the existence of living beings on our planet earth. The scientists look for water in space exploration in the view that if water can be traced on a planet, life can also be found. The quantity of water created at the time of the formation of earth, remains constant till to-date; however, water changes its location and form and thereby creates vital source of life and at the same time brings catastrophies. Management of water has many dimensions. Fristly, water needs to be available for drinking by human beings and other animals; and this form of water must satisfy quality standards. Secondly, water needs to be available to irrigated and natural crops, which provide basic food source to humans and animals. Thirdly, water is a source of recreation as evidenced by the facts that people love to go to water parks, rivers, reservoirs and ocean. Most expensive houses are often water-featured in many different forms. For these types of water bodies, one needs to maintain 'recreational values' of water. Fourthly, water can bring untold destruction/economic damage through floods, tsunami, droughts and landslides. Hence, management and conservation of water resources are given a high priority to a nation's vision and mission to achieve economic growth and sustainability.

The water problem arises from the point of 'quantity, quality, access, temporal and spatial distribution and affordibility'. There is plenty of water in ocean but this cannot be used for drinking (i.e. it does not meet quality criterion); however, this can be used for other purposes e.g. navigation and recreation. There might be lots of water during monsoon or rainy season but there may not be enough water during the dry season of the year (i.e. problem of temporal distribution). For example, over 90% of rainfall happens in Bangladesh during the 4 months period (rainy season), with no/little rain during the winter months. There might be enough water in an urban water supply but most of the poor people cannot access it as they live in temporary houses which do not have urban water supply, and also water is too expensive for most of them to buy (i.e. affordability problem).

Water management problem in Australia has a number of characteristics. Firstly, it has the most variable rainfalls in the world e.g. it suffers from long lasting droughts, followed by devastating floods. Australia is number one in the world in terms of per capita storage volume as it needs many reservoirs to capture and store water for supply during both wet and dry periods (Boughton, 1999). Secondly, Australia has one of the highest per capita water consumption althogh it is regarded as the most dry inhabitated continent on earth. Australia's 75% of interior areas are regarded as arid and semi-arid with mean annual rainfall smaller than 300 mm (Zaman et al., 2012).

Water conservation in Australia is high on national agenda. In urban areas where about 80% of Australian polulation live supply of adequate potable water is a challenge since water demand in Australia is very high and at the same it suffers from frequent droughts, which sometimes last for up to five years or even longer. Water infrastructure in large Australian cities are getting older, which may need a huge investment to maintain and replace. To cope with low water supply in the reservoir during droughts, a number of mechanisms have been proposed. Firstly, imposing a restriction on certain types of water uses (e.g. gardening, car washing and hard surface cleaning). Secondly, provision of alternative water supply such as rainwater harvesting from urban roofs. As reported by ABS (2010), about 26% of Australian households use a rainwater tank as a source of water.

Many studies have demonstrated the success of rainwater harvesting in Australia (e.g. Van der Sterren et al., 2012; Rahman et al., 2010; 2012a, b; Imteaz et al., 2011a, b; 2012). It has been reported by ABS in 2007 that the second most common source of water in Australian households is the grey water after mains water. About 54% of Australian households use grey water to source their water. About 93% of Australian households get water from mains (ABS, 2007). In case mains water is not available, people get their water from rainwater tanks, groundwater wells or other sources. Australian households often have some sorts of water conservation devices. As reported by the ABS, in 2007, 81% of households have a dual flush toilet, 55% dwellings have water-efficient shower heads. Also, community education and non-structural low cost measures such as water sensitive urban design have been playing a notable role in saving water in Australia. In addition to the above measures, water restrictions and increased water price are viewed as 'effective means' of reducing water demand. This chapter investigates how water price and demand restriction can assist in conserving water by reducing water demand. This presents a critical literature review, followed by a data analysis from the Blue Mountains Region in New South Wales, Australia to assess the impact of water restrictions on water demand.

WATER PRICE AND DEMAND: AN EVALUATION

Drinking water in Australia had been relatively cheaper or nearly free for long time. Both urban and rural consumers took for granted cheap or free clean water as a right which was mostly ensued by the scientific and engineering achievements and huge government investments on water sector. However, recently this complacency has been changed dramatically. It would not be overwhelming if water is named second oil in the context of many countries in the world. The price of water in major countries in the world is presented in Figure 1, which shows that water price in Australia is among the top 4 countries in the world.

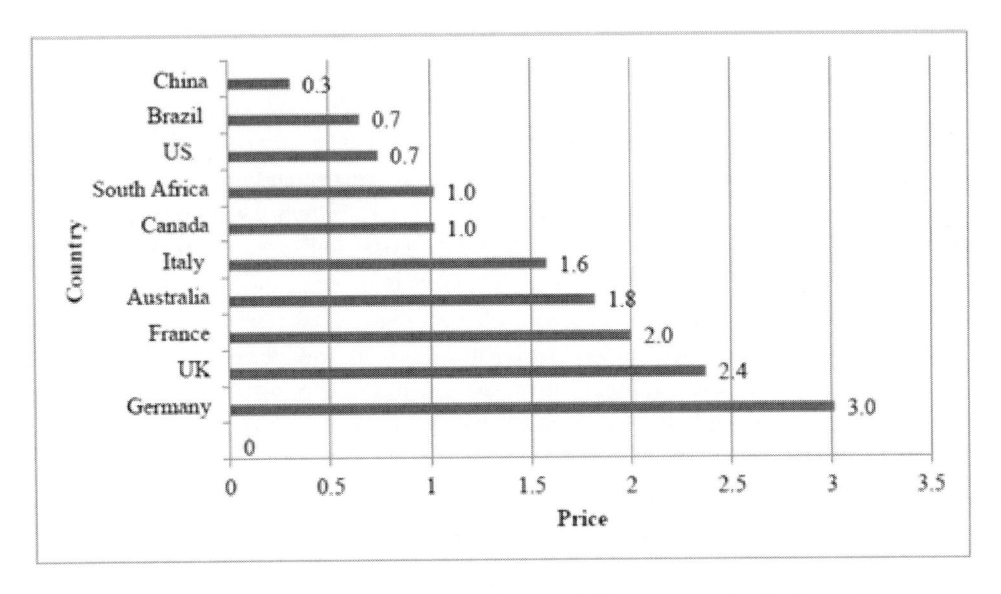

Figure 1. Price of water in US $ per cubic meter in 2009 in different countries in the world.

The main constrain of water scarcity in Australia is partly due to the high variability in rainfall, environmental issues and climate variability and change driven factors. One of the main reasons behind this scarcity in major Australian cities is a double digit population growth in these cities e.g. Sydney, Melbourne and Perth (Sibly, 2006). Under the current allocation arrangements by the government, city dwellers are tapped in danger of using supplementary water extracted from the surrounding environments. Thus, it is evident that demand and supply of water are negatively correlated (e.g. during high rainfall regime, irrigation demand is less) (Grafton et al., 2010). This reduced supply also coincides with greater demand of water in certain climatic condition. Other factors behind the supply inadequacy include inadequate investment in water transportation and infrastructure in last 20 years (Dwyer, 2006), restrictions on rural-urban water trade in terms of regulatory mechanism (Productive Commission, 2006) and urban water pricing that does not account for large differences in water supply (Grafton and Kompas, 2007). For example, in South Australia, all the consumers are responsible to pay same rate of water usage price for using potable water regardless of their locations though water supply cost is higher in the regional towns than in the metropolitan areas (Willis et al., 2013). As a result balancing between supply and demand has become a critical task in an unpredictable climate subject to extended periods of low rainfall, which is a common trend climate in Australia. The supply challenge is made worse by the substantial cost accounting for an excess of $2 billion and requires time at least 10 years upward under the conditions of normal rainfall to build and fill any new dam (New South Wales Government, 2004).

However, like many developed economy water usage price have been increasing in most of the capital cities in Australia and interestingly at a rate greater than inflation. Household water price have risen by 17% in 2010-11 financial year (ABS, 2012). According to Australian Bureau of Statistics report, the average price of water for households increased to $ 2.44/kL in 2010-11 from $ 2.10/kL in 2009-10. Table 1 represents a comparison of water consumption and water price between 2009-10 and 2010-11. As can be seen in Table 1, water price was increased around 25% though water consumption is somewhat in negative trend in between 2009-10 and 2010-11 period.

Moreover, recent investment will definitely force the urban water sector in the future to increase water usage price. WSAA (2010) estimated that water demand in six major cities in Australia would increase by 40–50% from 1505 GL per year in 2009 to approximately 2100 GL per year by 2026. This data provides an insight that Australian water sector will require new investment in water and wastewater infrastructures, which eventually place further pressure on water prices in coming days to recover the cost. Willis et al. (2013) mentioned in their study that reason for increasing water usage prices is more about recovering the full cost than it is about conserving water.

As an immediate response to supply concerns due to low reservoir level, the NSW government imposed water restrictions from October 2003. This restriction reduced overall water demand but still found insufficient to balance the gap between real demand and supply. In addition to this household were also subsidies to retrofit water-efficient products and installation of rainwater tanks on new dwellings designed to reduce water consumption by 40% compared to the current Sydney household average (NSW Government, 2004). In view to a positive and economically effective solution to current water crisis in Australia; researchers, policy planners and water supplying agencies suggested quite varying types of solutions for demand side management.

Table 1. Comparison of change in water consumption and price 2009-10 vs. 2010-11 (ABS, 2012)

	Year (2009-10)	Year (2010-11)	Percentage Change
Household Consumption (GL)	1844	1699	-8%
Agricultural Consumption (GL)	6987	7175	3%
Total Consumption	13515	13337	-1%
Household Water Price ($/kL)	2.1	2.44	17%
Agricultural Water Price ($/kL)	0.11	0.14	25%
Total Price ($/kL)	0.82	1.03	25%

These solutions can be categorized into two distinctive types, price measures and non-price measures. Interestingly, a considerable debate has been observed among the group of interested parties regarding the efficiency level of each types of regulatory system. This ongoing argument has embroiled water management issue more critical than ever. By supporting the price regime economists are generally in agreement that urban water prices that reflect marginal costs is a means of reducing demand of water specially during periods of inadequate water supply, while others argue that real demand for urban water is relatively price inelastic and price regime is an ineffective tool for regulating the imbalance between demand and consumption (Hoffman et al., 2006).

However, keeping the non-price measures aside, a variety of researches have been emerged in last three decades which specifically addressed different approaches to water pricing mechanisms not only in Australian context but also other parts of the world where water crisis is equally prevalent. It is revealed from the past literatures that a key feature of the policy about the demand side management is the pricing structures pertinent to water services. Study of the effects of pricing structure can be explained by the effectiveness of price in terms of regulating water consumption and meeting the multiple objectives usually taken into account while designing an optimal pricing policy. According to few literatures, invariably all pricing structures are complex as it attempts to meet the changeable and in most cases competing objectives like equity, financial stability, public tolerability, transparency and profitability.

It is also revealed from the reviewed literatures that for crafting an effective price mechanism for water comprehensive demand estimation is prerequisite. According to Worthington and Hoffman (2006) there are three salient features that need to be recognized before attempting to an effective pricing for water, first the presence of individual meter in all household; second the structure of prices representing the split between fixed and variable prices including any variance in these prices and third the billing frequency which indicates how often bills are issued to households for paying their water consumption. However, a variety of alternative pricing methods has been employed by researchers in the past years for attempting to meet these criteria (Dinar and Subramanian, 1998; Bartoszczuk and Nakamori, 2004). These range from a fixed charge invariant to the level of consumption; a fixed charge with a some free allowance; fixed charge + free allowance followed by some excess charge for consumption up to a certain level; a two-part tariff consisting of a fixed component (usually known as meter charge in many countries) and a usage component based on the actual amount of water consumed (usually known as a volumetric charge). The latter is being practiced commonly in Australia and other developed countries.

These prices per unit volume of water can be in three categories: (i) fixed block; (ii) increasing block, which increases with each successive block of water; (iii) decreasing block, which decreases with each successive block of water use. Massarutto (2007) reported that the cornerstone of European Union's Water Framework Directive is full pricing which has already been set. In order to facilitate appropriate water pricing, computer-assisted models can be useful. Such computer-assisted pricing are usually called "smart markets" (McCabe et al., 1991). Smart markets approach model has the computational intelligence of presenting real-time data known as "model verite" in agent-based computational economics.

Arbues et al. (2000) reported that Zaragoza, Spain practices the most complex block style tariff structure under the local water supply mechanism. It consists of 140 progressive pricing blocks with the total bill charged at the highest block price for the period. However, all these different types of pricing approaches suffer from arguments and counter arguments. One of the dominant suggestions obtained from most of the American and Australian studies is the introduction of household metering system (Yepes and Dianderas, 1996; Dalhuisen and Nijkamp, 2001; Dalhuisen et al., 2001; Bartoszczuk and Nakamori, 2004). It has been suggested in many research that the introduction of metering results in a reduction of water use regardless for all pricing structure used for water management. Yepes and Dianderas (1996) argue that the use of household metering can considerably reduce system maintenance efforts of water management. The researchers also added that unmetered water represents around 10% to 15% of the water supply in developed countries and more than 50% in developing countries. Whittington (1992) argues that increasing block rate tariffs may reduce welfare in developing nations. One reason behind his argument is fairly logical as he mentioned that generally multiunit apartments consist of a single meter but the combined use of multiple families in one apartment block increases the chance of the total billing amount against a meter reaching in the highest block prices. However, metering in developed countries is by no means universal (e.g. US, Canada, United Kingdom, Norway and Ireland).

In Australia many local councils and water supply authorities are still struggling hard with the issue of costs and benefits for installing individual household meter especially in apartments. Few literatures relate to the billing cycle claimed that for a given household and level of water consumption, billing frequency is inversely related to cost per assessment. The theoretical argument behind the frequency of billing cycle is that households can be more aware of the impact on income of large bills and these can potentially reduce water consumption in subsequent periods. It is also hypothesized that a frequent billing system will also remind the consumers more repeatedly about cost of their water consumption. Contrarily a less frequent billing would not permit households to adjust consumption very promptly if they encounter a sudden rise in water bill. Incidentally few models on billing frequency found statistically insignificant impact on water consumption (Griffin and Chang, 1990; Stevens et al., 1992). In Australia a 90-day billing cycle has been practiced which is corresponding to the quarterly rates assessment. However, generalization of current and past pricing regimes for urban water in Australia is quite difficult as each region is being operated by its own water provider under special institutional arrangements. Usually all urban water providers are an agency of either state or local government. This heterogeneity of institutional arrangements related with water providers leads to a variety of pricing methods as each authority act in response to local political and consumer demands (Sibley, 2006).

The NSW Independent Pricing and Regulatory Tribunal (IPART) set the maximum retail price for water in Sydney.

IPART stated preference was to specify water prices with reference to the long-run marginal cost of supply (LRMC) which it estimated between $1.20 and $1.50 per kilolitre (kL) in 2004/2005 (IPART, 2005). This prices was fixed for until 30 June 2009 and IPART also stated that water pricing should account for the'. . . imbalance between the demand for water and the available supply . . .'(IPART, 2005, p. 105 cited in Grafton and Kompas, 2007). At that time the Tribunal established a two tier increasing block pricing system where the higher Tier 2 price was imposed when households exceed 100 kL per quarter. However, a sharp increase in water usage has been observed. The annual bill of a single dwelling household with average water consumption of 200kL per year is increased by 2.6% less than inflation by the end of the 4-year determination period effective from 1 July 2012 to 30 June 2016. If the assumed inflation is to be 2.5% each year, then most households will experience modest increases in total water and sewerage bills but the bills will increase at a slower rate than other household items (IPART, 2012). Customers living in apartments in a multi-dwelling property usually share a common single meter. Sydney water used to charge customers who shared a meter a share of a meter based charge while those with individual meters pay the same standard charge as a separate house. On average those customers on shared meters usually pay around $70 per year. It was also reported that few household pay as little as $5 to $10 a year. This imbalance influenced IPART to introduce a standard water service charge for all residential customers. As a result, many households in units with a common meter are imposed to pay a higher water service charge than before. According to the report the usage charge for water in Sydney remains unchanged at $2.10 per kL (IPART, 2012). The current price regimes sets by IPART for other states in Australia can be seen in IPART official website www.ipart.nsw.gov.au.

Nevertheless, urban water lacks a history of efficient pricing. Urban water prices typically aimed to achieve only partial cost recovery but considering the rate of return on water assets it is still far below a commercial level. The debate on sustainable or efficient water pricing approach has emerged soon after the introduction of water restriction by many water authorities in 2003 in Australia. According to the reviewed literatures this water pricing debate is also prevalent in many developed and developing nations. Russell and Shin (1996) stated that efficient water price is the long-run marginal cost (LRMC) of supply in most cases. However he added that in some cases charging short-run marginal cost (SRMC) may be efficient too. LRMC represents the full economic cost of water supply.

In order to justify the dilemma of efficient pricing methods over non-price measures few notable studies can be mentioned. These analysts have estimated the economic losses from CAC (command-and-control) water conservation policies. Timmins (2003) compared a mandatory low-flow water appliance regulation with a modest increase in water price. The researcher used data from 13 - California cities running under groundwater treatment and found that price regime to be more cost effective than any technology standards for regulating water demand.

Brennan et al. (2007) concluded that the economic costs of utilising sprinkling restriction two days in a week in Perth are just below $100 per household in a particular season and the cost of a complete ban of outdoor watering ranges from $347 to $870 per household per in a specified season. Grafton and Ward (2008) reported that mandatory water restrictions in Sydney in 2004-2005 resulted in economic losses of 235 million dollars. Klawitter (2003) also suggested sustainable urban water pricing which must be designed to meet the needs of present and future generations.

He also emphasized on efficient use of resource, full cost recovery which must includes supply costs, opportunity costs and economic externalities, economic viability of the water utility and equity as well as fairness for different users. Olmstead and Stavins (2009) suggested that water pricing would be more cost effective in reducing water demand than the non-price means of demand management.

However, on the basis of few literatures discussed above it is somewhat imperative that if policymakers want to use prices as a tool for demand management then the key variable of interest will be the price elasticity of water demand. From this point of view it is hypothesized that an increase in the water price leads consumers to use less water, thus, price elasticity is a negative numeral. Olmstead and Stavins (2009) stated that important standard elasticity is - 1.0 which divides the threshold demand into two specific categories, elastic and inelastic demand. According to the authors there is a critical distinction between inelastic demand and demand which is unresponsive to price. The authors also hypothesized that if demand is really unresponsive to price then the price elasticity will be equal to zero and the demand curve will be a vertical line. This suggests that same quantity of water will be demanded at any price. This is somewhat practical for a subsistence quantity of drinking water. However, the authors also reported on the basis of their empirical econonomic analysis which includes a 50 years calculation about the trends of urban water usages that this is not practical. Some literatures also reported that the price elasticity of residential water demand varies across place and time. However one study mentioned that on an average in the United States a 10% increase in the marginal price of water in the residential water in urban area can be predictable to lessen demand by about 3-4% in the short term. This result is also consistent with results reported by two notable studies attempted to empirical estimations of the price elasticity of residential energy demand by Bohi and Zimmerman (1984) as well as Bernstein and Griffin (2005). According to the authors if a water utility wanted to reduce demand by 20% (especially during relatively low rainfall or a drought) considering an elasticity of - 0.4, it will require approximately a 50% increase in the marginal water price. Last but not the least price elasticity varies with many other factors. It is also found that in the residential water sector high-income groups tend to be less sensitive to water price increase than the low-income households.

Despite the theoretical debates alongside regarding various price approaches attempted to address as regulatory regimes and their efficacy to reduce water usages, one notable study has been published by Sydney Water in 2011 about the price elasticity of water demand for residential use. This gives some sort of evidence that elasticity of water demand is price sensitive but its impact on the water price is quite low in the short run. This study was conducted to estimate the responsiveness of three different types of consumers, owner occupied houses, tenant occupied houses and apartment units to analyze the impact of increases in water usage prices. At the same time it investigates the likely impact of any future changes to water price as well. This study utilized 'panel data' analysis by incorporating a sample of roughly 95,000 individual households and 3,300 blocks of apartment units. A remarkable feature of the study is that it applied a dynamic model specification by using a functional form which allowed respondents to be more sensitive to water usage prices, the higher the level of prices. This particular research found that at a water usage price of $1.20 per kL (estimated in $2009-10 dollars) the projected immediate and long-term real price elasticities for the demand for water ranges from -0.01 to -0.08 and -0.03 to -0.14, respectively.

They found that an increase in water price by 10 % would result in reduction of residential water demand by 0.5% immediately. In the following one year, water demand would be reduced by a further 0.6 %. From various viewpoints it is somewhat assumed that water usage price as a means of demand management might be encouraging provided that an efficient pricing mechanism is to be in place considering welfare cost.

DEMAND RESTRICTIONS VS WATER DEMAND

Urban water planners and policy makers face many challenges to secure proper water supply for the people during the period of droughts. In Australian capital cities, about 50% of residential water is used on private outdoor areas (Brennan et al., 2007). As a result, use of water restriction to effectively manage water usage on private outdoor areas during the drought periods has become a popular method to many water authorities in Australia. As examples, due to recent prolonged droughts in Brisbane, Melbourne and Sydney, the three major cities in Australia, water authorities imposed water use restrictions of varying severity to their customers to reduce water demand as the dam water storage levels dropped quite low (Herriman et al., 2009; Queensland Water Commission, 2010; Sydney Water, 2010). Moreover, many studies on climate change issues have predicted increased drought frequency and related low water availability in different locations such as South-eastern Australia (Hennessey et al., 2007). This will increase the challenges to manage inadequate water supply. Therefore water restriction policy may become a popular means of managing short water supply during the emergency periods in the future.

Water restrictions can be imposed in several forms. They are typically implemented by specifying the time of day for garden watering, the maximum length of the watering period and allowances for hand watering (MacDonald et al., 2010). Prohibition for using hoses to wash hard surfaces, restrictions on car washing and filling or refilling of swimming pools are some of the common forms of these restrictions. For example, hosing of lawns and gardens were not permitted in any time during the Level 2 water restrictions in Sydney. Only hand-held hosing was permitted before 10 am and after 4 pm on Wednesday, Friday and Sunday. The severity and timing of water restrictions are largely determined based on the dam storage levels. Such as in Sydney, Level 1, Level 2 and Level 3 water restrictions were imposed when the dam levels dropped below 60%, 50% and 40%, respectively (Spaninks, 2010; Sydney Water, n.d). Level 1 and Level 3 were the most liberal and most stringent water restriction, respectively. Description and introduction time of different levels of water restriction in Sydney are provided in Table 2.

Water restrictions are normally imposed as a short-term demand management policy to manage inadequate water availability/supply during the emergency times (e.g. drought periods) and to avoid or delay the costly expansion of the water supply system. However, water restrictions in different Australian cities have been implemented for long term period. For example, mandatory water restrictions were imposed on Sydney residents in October, 2003 and it was withdrawn in June, 2009, after about six years. As Australia is characterised by extreme diversity between locations, extensive differences exist in the restriction rules, implementation timing, and severity levels on residential and non-residential sectors across Australia.

Table 2. Levels, scope and timing of water restrictions imposed in Sydney

Restriction Rules	Restriction rules belongs to restrictions levels	Introduction date
I. No hosing of hard surfaces and vehicles.	Level 1 (I +II)	1-Oct-03
II. No use of sprinklers or other watering systems.	Level 2 (I + II + III + IV)	1-Jun-04
III. No hosing of lawns and gardens, only hand-held hosing was allowed for three days in a week (before 10 am and after 4 pm on Wednesdays, Fridays and Sundays.)	Level 3 (I + II + III + IV + V +VI)	1-Jun-05
IV. No filling of new or renovated pools over 10,000 L except with a permit from Sydney Water.		
V. No hosing of lawns and gardens, only hand-held hosing was allowed for two days in a week (before 10 am and after 4 pm on Wednesday and Sunday.)		
VI. Fire hoses used only for fire fighting purpose and not for cleaning.		

Different rules have been applied to different cities based on the nature of the water resources availability. Moreover, naming conventions are different in different cities, such as for Sydney the restrictions have been named as “levels” whereas for the Melbourne, these have been named as “stages”.

In addition, some base level restrictions have become permanent in some Australian cities. Such as ‘Water Wise Rules (WWR)’ (i.e. all hoses must have a trigger nozzle, garden watering is allowed before 10 am and after 4 pm only) in Sydney and permanent water use rules (e.g. watering systems such as manual, automatic, spray or dripper are allowed to use in the gardens and lawns from 6 pm to 10 am) in Melbourne have been put in place since 21 June, 2009 and 1 December, 2012, respectively.

As the dam storage levels are in good condition for the last couple of years due to recent La-Nino dominated climatic regime, mandatory water restrictions have been withdrawn in many metropolitan water areas. These restrictions rules have been replaced by permanent water conservation measures that reflect basic water-efficient practices such watering the garden in the night time only.

These permanent conservation rules are known as WWR in Sydney, Permanent Water Savings Rules (PWSR) in Melbourne, Permanent Water Conservation Measures (PWCM) in Australian Capital Territory (ACT) and Water Wise Measures (WWM) in Adelaide. Restriction periods and levels of severity of imposed water restrictions of different Australian cities are summarised in Table 3.

Water savings from water restrictions are influenced by many factors such as design of rules, severity level of water restrictions, promotion, education, community awareness and enforcement activities. In addition, some external factors such as climate and availability of other water sources also influence the savings from water restrictions.

Table 3. Water restrictions periods and severity levels of major Australian cities

Australian Cities	Restriction periods	Restriction Levels/stages	Current Restrictions (2013)
Adelaide	October, 06 to November, 10	Stage 2- Stage 3	WWM
Melbourne	November,02-November, 12	Stage 1- Stage 3	PWSR
Sydney	October, 03 - June, 09	Level 1 - Level 3	WWR
Brisbane	May, 05 - December, 12	Level 1 - Level 5	No Restrictions
Perth	November, 94 - continue	Stage 1- Stage 4	Stage 1
ACT	December, 02-October,10	Stage 1- Stage 3	PWCM

By quantifying the savings from water restrictions, effectiveness of these programs can be evaluated which can help to plan and design future water demand management strategies for a given water supply system. Water savings from water restrictions can be estimated by several ways:

1. End-use approach
2. Long term average method
3. Before and after method
4. Climate corrected method

End use approach involves collecting information and data about the effects of specific water restriction rules on residential and non-residential water use behaviours. Such as, if the data can be collected about the volume of water used for gardening under the sprinkler ban and the volume of water without sprinkler ban for a week, then these volumes can be compared each other to evaluate the effectiveness of this particular restriction rule. However, this approach is time consuming and costly as estimating water savings to specific restriction rules or individual restrictions levels require time series surveying data and metering of a representative sample of households in each location.

Using long term average method, water savings can be estimated by comparing daily/ monthly metered water use during the restriction period with the historical average of daily/ monthly water use.

Here historical or base water use period need to be defined first to get the average long term water usage for at least 3 to 5 years under no restriction circumstances. For an example, base water use periods can be taken from the years before 2003 for Sydney region, as there was no mandatory water restrictions imposed before 2003. Mandatory water restrictions were in place in Sydney from the period of October, 2003 to June, 2009.

Water savings through long term average method can be estimated by the following equation:

$$(W_s)_i = (\overline{W}_u)_b - (W_u)_i \quad (1)$$

where, $(W_s)_i$ = water savings for any day/month, $(\overline{W}_u)_b$= average water usage for the base period (3 to 5 years) in daily/monthly and $(W_u)_i$ = water use for any day/month under restriction conditions.

For an example, calculation of water savings of the year 2004 for the single dwelling residential sector in the Blue Mountains Water Supply System (BMWSS), NSW, Australia, is presented in Table 4. 'Before and after method' consists of comparing daily/monthly metered water usage under restriction period with the average water usage of that particular day or month in the previous years under no restriction conditions. For an example, estimation of water savings for the year of 2005 of the single dwelling residential sector in the BMWSS, NSW, Australia is presented in Table 5.

As per climate corrected method, expected water use is estimated/modelled first that would have occurred due to climatic condition and under no water restriction for a particular day/month that is originally under restriction condition.

Thereafter, estimated/modelled water demand value is compared with the observed water demand value under restriction condition for a particular day/month to get the value of water savings. Water savings through climate corrected method can be estimated by the following equation:

$$(W_s)_i = (W_e)_i - (W_u)_i \tag{2}$$

where, $(W_s)_i$ = water savings of a day/month, $(W_e)_i$ = expected (modelled) water use under assumed no restriction condition, $(W_u)_i$ = water usage of a day/month under restriction condition.

For an example, calculation of total water savings of the year 2006 by climate corrected method for the single dwelling residential sector in the BMWSS, NSW, Australia is presented in Table 6.

Water savings from different levels of water restriction by long term average method is presented in Figure 2 for the BMWSS, NSW, Australia.

Table 4. Calculation of total water savings by long term average method

Year	Month	Total Consumption (kL)	Number of Dwellings	Per Dwelling Consumption (kL/dwelling/ month) (a)	Average Base Water Consumption (1997-2002) (kL/dwelling/ month) (b)	Total Water Savings (kL/dwelling/ month) (b-a)
2004	1	248876	16156	15.40	16.73	1.33
2004	2	230598	16159	14.27		2.46
2004	3	232405	16157	14.38		2.35
2004	4	224835	16161	13.91		2.82
2004	5	227949	16165	14.10		2.63
2004	6	212968	16166	13.17		3.56
2004	7	219926	16180	13.59		3.14
2004	8	219835	16199	13.57		3.16
2004	9	214196	16215	13.21		3.52
2004	10	221542	16237	13.64		3.09
2004	11	214650	16245	13.21		3.52
2004	12	216043	16260	13.29		3.44

Table 5. Calculation of total water savings by 'Before and After Method'

Year	Month	Per Dwelling Consumption (kL/dwelling/month) (a)	Monthly Base Water Use (kL/dwelling/month) (b)	Total Water Savings (kL/dwelling/month) (b-a)
2005	1	13.11	19.36	6.25
2005	2	11.83	16.26	4.42
2005	3	13.02	16.01	2.99
2005	4	12.71	15.38	2.67
2005	5	13.10	14.87	1.77
2005	6	12.43	14.09	1.65
2005	7	12.78	14.72	1.94
2005	8	12.78	17.33	4.55
2005	9	12.29	17.57	5.29
2005	10	12.63	18.30	5.67
2005	11	12.24	18.42	6.19
2005	12	13.52	19.50	5.98

Table 6. Calculation of water savings by climate corrected method

Year	Month	Observed Per Dwelling Consumption (kL/dwelling/month) (a)	Expected Per Dwelling Consumption (kL/dwelling/month) (b)	Total Water Savings (kL/dwelling/month) (b-a)
2005	Jan.	13.11	17.56	4.45
2005	Feb.	11.83	16.46	4.63
2005	Mar.	13.02	16.21	3.19
2005	Apr.	12.71	15.26	2.55
2005	May	13.10	14.16	1.06
2005	Jun.	12.43	12.27	0.00
2005	Jul.	12.78	13.76	0.99
2005	Aug.	12.78	17.65	4.87
2005	Sep.	12.29	16.31	4.02
2005	Oct.	12.63	16.85	4.22
2005	Nov.	12.24	15.90	3.66
2005	Dec.	13.52	19.36	5.83

As can be seen in Figure 2, water savings during the Levels 1, 2 and 3 water restriction periods were found to be 9%, 18% and 20%, respectively for the single dwelling residential sector. Similar results were reported by different water authorities in different Australian cities, such as Sydney Water reported 13%, 16% and 17% savings for Sydney under Level 1, Level 2 and Level 3 water restrictions, respectively.

Yarra Valley Water reported 8% and 11% water savings for Melbourne under Stage 1 and Stage 2 water restriction, respectively. ACTEW Water reported 11% to 33% water savings under different stages of water restrictions.

These findings indicate that the water restrictions were able to reduce water demand during the drought periods in Australia.

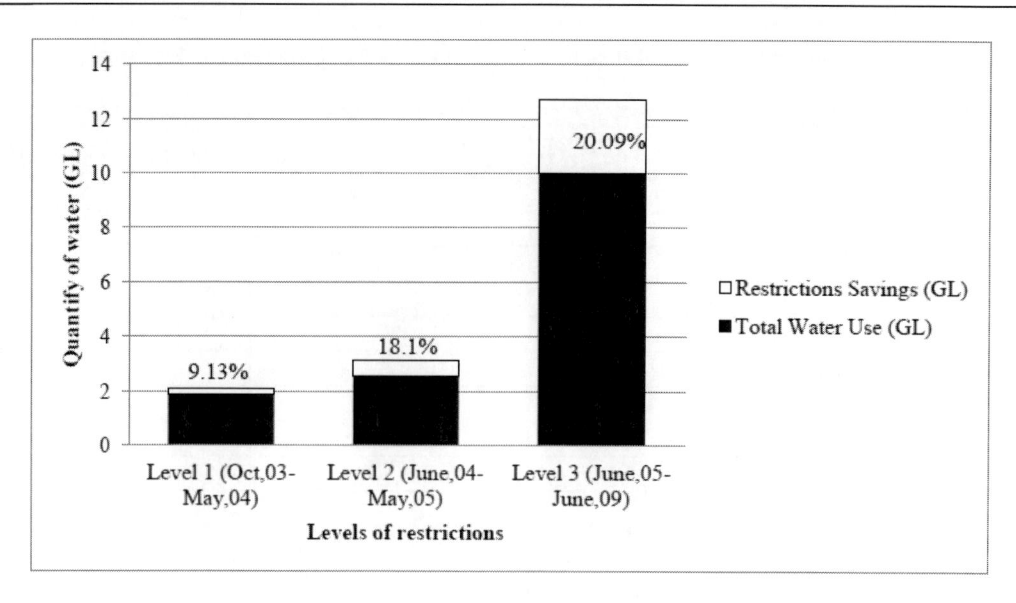

Figure 2. Water savings from different levels of water restriction in the Blue Mountains Water Supply Systems, NSW, Australia, during drought periods.

WATER PRICE AND WATER RESTRICTION: EFFECTIVENESS IN AUSTRALIAN PERSPECTIVES

Two major options normally exist for the urban water planners/water authorities to manage water scarcity which is either supply side and/or demand side management. Supply side management generally involves in finding new water sources which include building new infrastructure (e.g. desalination plants, dams and reservoirs), capacity expansion of existing system and sourcing new catchments. On the other hand, demand side management involves implementing various regulatory policies and conservation measures to reduce water consumption. The demand management mechanisms are normally either price based or non-price based. Public education, campaign, water efficiency products and water restrictions are some of the forms of non-price instruments. In Australia, effectiveness of water price mechanism to regulate water demand is still in debate. While economists generally agree that urban water price can be an effective tool to reduce demands during the period of water shortages, others argue that price elasticity of demand is very low (i.e. large changes in price is associated with a small change in water demand). The productivity commission (PC, 2011) has made a recent public inquiry into Australia's Urban Water Sector and reported that use of scarcity pricing are not as effective as other policy changes (i.e. water restrictions). This is because water demands have been found to be mostly insensitive to price in the short run. Since the effectiveness of using water price as a means of balancing the supply and demand is mainly dependent on the short run price elasticity, it could be said that water price mechanism might not be an effective tool in reducing water demand during the drought periods.

On the other hand, non-price mechanisms such as water restriction have been widespread in Australia for long periods to manage short water supply during droughts condition. It has been found that water restrictions were capable of reducing water demand during the emergency period up to 33%.

Therefore, it has demonstrated to be an effective way of achieving short term reduction in water consumption during the emergency periods. However, water restrictions have seemed to impose cost on households and affected the community in negative ways to some extent (e.g. decrease consumer welfare, loss of amenity, recreational values from gardens and pools, additional travel times to recreational places, financial costs or additional time required for changing private gardens or watering technologies). In addition, few Australian studies have addressed the question of willingness to pay (WTP) to avoid water restrictions. For example, Gordon et al. (2001) found that consumers were willing to pay an extra $150 per annum to avoid mandatory water restrictions. Hensher et al. (2006) found that households were willing to pay up to $239 extra on their water bills to remove sprinkler restrictions. These two notable studies indicate that water consumers might be less reactive to higher water price than a restriction. Therefore, water restrictions should be assessed against other available options such as capacity expansion, large scale water recycling and reuse, alternative water resources and other incentive mechanisms in both long-term and drought-response planning.

Conclusion

In Australia, water usage prices have been found to be nearly inelastic in most of the cases which indicate that a large increase in the water price would be necessary to get a significant immediate reduction in water consumption. However, water demand seems to be more sensitive in the long run which may lead to use water price mechanism as a means of reducing water consumptions. On the contrary, water restrictions have played an important role in Australia as a drought response option to reduce water consumption during drought period. It has been found that water restrictions are capable of reducing water demand during the restriction period up to 33%. This was successful in reducing water consumption and in relieving pressure on declining storage levels during the drought periods. Therefore, water restrictions could be an alternative choice in the long term planning and drought response options. However, as these restrictions have some negative effects on the community, this option need to be carefully investigated in terms of cost and benefits against some other alternatives such as water price and supply augmentation to ensure proper water supply to the people.

Acknowledgments

Water consumption data of the Blue Mountains Water Supply System were obtained from Sydney Water. The best available data at the time of study has been used, which may be updated in near future. The authors express their sincere thanks to Pei Tillman and Frank Spaninks of Sydney Water for their assistance in collating and providing the data. Further, the authors are grateful to Lucinda Maunsell of Sydney Water, and Mahes Maheswaran and Golam Kibria of Sydney Catchment Authority for their cooperation and assistance during data collation and analysis. The authors also wish to thank Dr Dharma Hagare of University of Western Sydney for his thoughtful comments on the water price and demand restriction issues.

References

Australian Bureau of Statistics (ABS). (2007). Environmental Issues: People's Views and Practices, Available at http://www.abs.gov.au.

ABS (2010). More Australian using rainwater tanks. Media Release, 19 Nov. 2010 by ABS. http://www.abs.gov.au

ABS (2012). Water account Australia 2010–11, cat. 4610.0, Canberra.

Arbués, F., Barberan, R., Villanua, I. (2000). Water price impact on residential water demand in the city of Zaragoza: A dynamic panel data approach. *Paper presented at the 40th European Congress of the European Regional Studies Association in Barcelona*, 30-31 August.

Bartoszczuk, P., Nakamori, Y. (2004). Modelling sustainable water prices. In: M. Quaddus and A. Siddique (eds.) *Handbook of Sustainable Development Planning: Studies in Modelling and Decision Support*. Cheltenham: Edward Elgar.

Bernstein, M. A., J. Griffin. (2005). *Regional differences in the price elasticity of demand for energy, technical report*, RAND, Santa Monica, Calif.

Bohi, D. R., Zimmerman, M. B. (1984). An update on econometric studies of energy demand behavior, *Annual Review of Energy*, 9, 105 – 154.

Boughton, W. C. (1999). *A century of water resources development in Australia*, 1900-1999, Institution of Engineers Australia.

Brennan, D., Tapsuwan, S., Ingram, G. (2007). The welfare costs of urban outdoor water restrictions. *Australian Journal of Agricultural and Resource Economics*, 51, 243-261.

Dalhuisen, J., Nijkamp, P. (2001). The economics of H_2O. In: *Conference Proceedings: Economic Instruments and Water Policies in central and Eastern Europe: Issues and Options*, Szetendre, Hungary, <http://www.rec.org/. > (Viewed on May.6, 2013).

Dalhuisen, J., de Groot, H., Nijkamp, P. (2001). Thematic report on the economics of water in metropolitan areas, European Commission: Environment and Climate Programme, DG XII, Human Dimension of Environmental Change: Metropolitan Areas and Sustainable Use of Water. <http://www.feweb.vu.nl/.> (Viewed on May.6, 2013).

Dinar, A., Subramanian, A. (1998). Policy implications from water pricing in various countries. *Water Policy,*1, 239–250.

Dwyer, T. (2006). Urban water policy: in need of economics, *Agenda,* 13, 3–16.

Gordon, J., Chapman, R., Blamey, R. (2001). Assessing the options for the Canberra water supply: an application of choice modelling, In: Bennett, J. and Blamey, R. (eds), *The Choice Modelling Approach to Environmental Valuation*. Cheltenham, UK, Edward Elgar 73–92.

Grafton, Q. (2010). Submission to the Productivity Commission in relation to its issues paper on Australia's urban water sector, *Australia's Urban Water Sector-Public Inquiry, Submission no. 22*, Productivity Commission, Canberra.

Grafton, Q., Kompas, T. (2007). Pricing Sydney Water. *Australian Journal of Agricultural and Resource Economics,* 51(3), 227–241.

Grafton, Q., Ward, M. (2008), Prices versus rationing: Marshallian surplus and mandatory water restrictions. *Economic Record*, 84, 57–65.

Griffin, R., Chang, C. (1990). Pre-test analyses of water demand in thirty communities. *Water Resources Research,* 26, 2251–2255.

Hennessy, K., Fitzharris, B., Bates, B. C., Harvey, N., Howden, S. M., Hughes, L., Salinger, J., Warrick, R. (2007). Australia and New Zealand. In: Parry, M., Canziani, O., Palutikof, J., van der Linden, P., Hanson, C. (Eds.), *Climate Change 2007: Impacts, Adaption and Vulnerability.* Contribution of working group II to the fourth assessment report of the intergovermental panel on climate change. Cambridge University Press. Cambridge, UK, 507-540.

Hensher, D., Shore, N., Train, K. (2006). Water supply security and willingness to pay to avoid drought restrictions, *Economic Record,* 82, 56–66.

Herriman, J., Chong, J., Campbell, D., White, S. (2009). *Review of water restrictions, Final report*, Volume 1-Review and Analysis. Institute for Sustainable Futures.

Hoffmann, M., Worthington, A., Higgs, H. (2006). Urban water demand with fixed volumetric charging in a large municipality: the case of Brisbane, Australia. *Australian Journal of Agricultural and Resource Economics,* 50, 347–359.

Imteaz, M. A., Rahman, A., Ahsan, A. (2012). Reliability analysis of rainwater tanks: A comparison between South-East and Central Melbourne. *Resources, Conservation and Recycling*, 66 (2012), 1-7.

Imteaz, M. A., Ahsan, A., Naser, J., Rahman, A. (2011a). Reliability analysis of rainwater tanks in Melbourne using daily water balance model. *Resources, Conservation and Recycling*, 56, 80-86.

Imteaz, M. A., Shanableh, A., Rahman, A., Ahsan, A. (2011b). Optimisation of rainwater tank design from large roofs: A case study in Melbourne, Australia. *Resources, Conservation and Recycling*. 55, 1022-1029.

Independent Pricing and Regulatory Tribunal (2005). *Sydney Water Corporation, Hunter Water Corporation, Sydney Catchment Authority Prices of Water Supply Wastewater and Stormwater Services,* Final Report.

Independent Pricing and Regulatory Tribunal (2012). Sydney Water Corporation, Sydney Water's price forcustomers in Sydney from 1 July 2012, Final Report.

Klawitter, S. (2003). A methodical approach for multi criteria sustainability assessment of water pricing in urban areas. *Paper presented at the 2003 Berlin Conference on the Human Dimensions of Global EnvironmentalChange.* < http://www.fu-berlin.de/.> (Viewed on May.9, 2013).

MacDonald, D. H., Crossman, N. D., Mahmoudi, P., Taylor, L. O., Summers, D. M., Boxall, P. C. (2010). The value of public and private green spaces under water restrictions. *Landscape and Urban Plannning*. 95(4), 192-200.

McCabe, K. A., Rassenti, S. J., Smith, V. L. (1991). Smart Computer-Assisted Markets. *Science*, 254, 534-538.

New South Wales Government (2004). Meeting the Challenges: Securing Sydney's Water Future. < http://www.dipnr.nsw.gov.au> (Viewed on Apr.12, 2013).

New South Wales Government (2006). Metropolitan Water Plan. <http://www.waterforlife.nsw.gov.au/about_water_for_life/metropolitan_water_plan> (Viewed on Apr.12, 2013).

Olmstead, S. M., Stavins, R. N. (2009). Comparing price and nonprice approach to urban water. *Water Resources Research*, 45, W04301, doi: 10.1029/2008WR007227.

Productivity Commission (PC) (2006). Rural Water Use and the Environment: The Role of Market Mechanisms. Research Report, Melbourne.

Productivity Commission (PC). (2011). *Australia's Urban Water Sector, Draft Report,* Canberra. Available in www.pc.gov.au.

Queensland Water Comission. (2010). South east Queensland water strategy. Queensland Water Commission, Brisbane, Queensland.

Rahman, A., Dbais, J., Imteaz, M. A. (2010). Sustainability of RWHSs in Multistorey Residential Buildings. *American Journal of Applied Science*, 1(3), 889-898.

Rahman, A., Dbais, J., Islam, S. K., Eroksuz, E., Haddad, K. (2012a). Rainwater harvesting in large residential buildings in Australia. In: "*Urban Development*", ISBN 978-953-307-957-8, Editor: Dr. Serafeim Polyzos, Published by InTech, www.intechweb.org, 21pp.

Rahman, A., Keane, J., Imteaz, M. A. (2012b). Rainwater harvesting in Greater Sydney: Water savings, reliability and economic benefits, *Resources, Conservation and Recycling*, 61, 16-21.

Russell, C., Shin, B. (1996). An application and evaluation of competing marginal cost pricing approximations, in Marginal Cost Rate Design and Wholesale Water Markets. *Advances Economics Environmental. Resources*, vol. 1, edited by D. Hall, pp. 141– 164, JAI Press, Greenwich, Conn.

Sibly, H. (2006). Efficient urban water pricing, *Australian Economic Review* 39, 227–237.

Stevens, T. H., Miller, J., Willis, C. (1992). Effect of price structure on residential water demand. *WaterResources Bulletin,* 28, 681–685.

Spaninks, F. (2010). Estimating the savings from water restrictions in Sydney. *Water*, 37(5), 65-69.

Sydney Water (2006). Mandatory water restrictions. <http://www.sydneywater.com.au/SavingWater/WaterRestrictions/> (Viewed on Apr.2, 2013)

Sydney Water. (2010). Water conservation and recycling implementation report, 2009-10. *Sydney Water*, New South Wales, Australia.

Sydney Water (2011). The residential price elasticity of demand for water. A joint study by Sydney Water and Dr Vasilis Sarafidis, Lecturer in Econometrics, University of Sydney.

Sydney Water (n.d.). “The history of water wise rules.” <http://www.sydneywater.com.au/water4life/waterwise/WhenWereWaterRestrictionsIntr oduced.cfm > (Viewed on Apr. 10, 2012).

Timmins, C. (2003). Demand-side technology standards under inefficient pricing regimes: Are they effective water conservation tools in the long run?. *Environmental and Resource Economics*, 26, 107-124.

Van der Sterren, M., Rahman, A., Dennis, G. (2012). Rainwater harvesting systems in Australia. In: *Water Quality*, ISBN 979-953-307-638-5, Editor: Dr Kostas Voudouris, Published by InTech, www.intechweb.org, 26 pp.

Willis, E., Pearce, M., Mamerow, L., Jorgensen, B., Martin, J. (2013). Perceptions of water pricing during a drought: A case study from South Australia. *Water,* 5,197-223.

Whittington, D. (1992). Possible adverse effects of increasing block water tariffs in developing countries. *Economic Development and Cultural Change,* 41, 75–87.

Worthington, A. C., Hoffmann, M. (2006). A state of the art review of residential water demand modelling.< http://ro.uow.edu.au/commpapers/301> (Viewed on Apr.5, 2013).

WSAA 2010, *Submission to Productivity Commission issues paper: Australia's Urban Water Sector,* Australia’s Urban Water Sector – Public Inquiry, Submission no. 29, Productivity Commission, Canberra.

Yepes, G., Dianderas, A. (1996). *Water and Wastewater Utilities Indicators.* 2nd Ed. Water and Sanitation Division, World Bank. <http://www.worldbank.org/.> (Viewed on Apr.12, 2013).

Zaman, M., Rahman, A., Haddad, K. (2012). Regional flood frequency analysis in arid regions: A case study for Australia. *Journal of Hydrology*, 475, 74-83.

In: Water Conservation
Editor: Monzur A. Imteaz
ISBN: 978-1-62808-993-6

Chapter 9

IMPACTS OF LAND USE AND LAND COVER CHANGES ON A MOUNTAINOUS LAKE

Hassen M. Yesuf[1*], Tena Alamirew[2], Mohammed Assen[3] and Assefa M. Melesse[4]

[1]Wollo University, Dessie, Ethiopia
[2]Haramaya University, Dire Dawa, Ethiopia
[3]Addis Ababa University, Addis Ababa, Ethiopia
[4]Florida International University, Miami, FL, US

ABSTRACT

Land cover assessment in a lake drainage basin is used to monitor changes in magnitude of physical parameters of a lake, including water balance components, rate of sediment deposition and morphometric parameters as a revelation of an intimate link between what happens in the lake with what is happening in its drainage basin. The objective of this chapter was to quantify the spatio-temporal changes in land use and land cover in the Lake Hardibo closed drainage basin occurred for the last 50 years using multi-temporal remote sensing geospatial data. Two time events of aerial photographs (1957 and 1986) and one satellite image (2007) historical land use and cover data were analyzed using combined geospatial computer software of ERDAS Imagine 9.2 and its core module Leica Photogrammetric Suite (LPS 9.2), and ArcGIS 9.3. An aerial coverage of land use and cover categories were identified and mapped for three historical time series. The catchment has seen significant land use changes over the last 50 years. The results showed that between 1957 and 2007 farmlands/settlements and shrublands/degraded lands have increased by 40.68 % and 50.91 % at the rate of 11.92 and 4.57 ha per annum, respectively. The grasslands, bushlands, forestlands and lake surface area diminished by 45.47 %, 51.31 %, 9.65 %, and 6.47 % at the rate of 1.42, 12.68, 0.43, and 2.19 ha per annum, respectively in the last half century. These changes might have brought significant consequences in the hydrological, soil erosion and sediment processes in the lake's drainage basin system.

* Email: hassen.mohammed@yahoo.com

Keywords: Lake Hardibo; land use/land cover change; LPS (Laica Photogammetric Suite); remote sensing; Ethiopia

INTRODUCTION

In contrast to many other environments, mountain ecosystems form mosaics of variability and have significant effects on the surrounding lowlands. When we intervene in these fragile environments, we take from the Gods an important part of responsibility for the reasonable use of mountain resources and for the sustainable development of mountain ecosystems. Research and development must therefore take account of this variability (Messerli et al. 1988). Hydrologic modeling of a watershed requires unique distribution of land use and land covers to simulate hydrologic, sediment and water quality parameters. Land use change is one of the most conspicuous changes in cultural landscapes in many regions of the world (Bormann et al. 2007). In addition to climate change, land use change has strong impact on the water budget of lakes and their drainage basins. The prediction of water fluxes in changing environments, under changing boundary conditions is an important task, and requires the use of hydrological catchment models.

Land use/land cover changes are important ingredients in the larger problem of environmental change. Land use patterns, driven by a variety of social and economical processes, result in land cover changes that affect biodiversity, water balance, reservoir/lake sedimentation, radiation budget and other factors (Riebsame et al. 1994).

Mapping current land use patterns is much easier job as compare to historic land use pattern. Advancement in remote sensing and geographical information system techniques have made it easier to collect geospatial information from multi-temporal aerial photographs and satellite images of most recent and earliest patterns of land uses and land covers (Tekle and Hedlund 2000; Zeleke and Hurni 2001; Bewket 2002; Ghaffar 2005; Akililu et al. 2007). Population increase was substantial in the highlands of Ethiopia during the 20th century. This resulted in increased deforestation, and intensified cultivation, as well as increased and accelerating soil and land degradation throughout the highlands. In addition, this had considerable effects on volume and velocity of surface runoff, which might have increased with intensified land use and accelerated land degradation during the same period (Hurni 1987; Hurni et al. 2005).

Pollen and charcoal analysis made by Darbyshire et al. (2003) on sediment cores taken from Lake Hayq and Hardibo provided evidence that the vegetation of the lakes' drainage basin had changed due to human influence during the last 3000 years. The natural climatic climax vegetation was *Podocarpus-Juniperus* forest. At about 500 BC, the forests were cleared and replaced by secondary vegetation of *Dodonea* scrub and grassland persisted for 1800 years. *Juniperus* forest, *with Olea and Celtis*, then expanded from AD 1400 to 1700, possibly because of drought-induced depopulation followed by increased rainfall. The authors further reported that a second phase of deforestation due to rising human populations and intensified land use had again increased during the last three centuries.

Lake Hardibo is an attractive highland mountain fresh water lake in the northeastern landscape of Ethiopia. The total dissolved solids (TSD) concentration of the lake is less than 780 mg/liter (Demlie et al. 2007). The lake has great significance to the riparian inhabitants by providing protein from fisheries, fresh water for drinking, preservation of sound ecosystems and biodiversity, and offers important recreational values. Despite its importance its drainage basin, especially the land surrounding the lake, is intensively cultivated contributing excessive sediment laden runoff into the lake during rainy seasons and drying up of stream flow in dry seasons; which invariably degrades the quality and quantity of the lake water and ecological integrity.

Historic land use data offers a unique opportunity to study the impacts of actual land use/cover changes on erosion processes, sediment discharges and water balance components, which are essential for the protection of the environment and ecosystems. The changing land use and cover patterns in Lake Hardibo drainage basin, driven by a variety of social and economic factors are affecting the lake and its basin biodiversity equilibrium, water balance, and sedimentation. The study of land use/land cover changes in the lake's drainage basin is an important ingredient in the environmental interpretation of the lake and will help develop an integrated watershed management practices in the future. Therefore, the objective of this study is to analyze land use/cover changes that occurred in Lake Hardibo drainage basin in the past five decades.

MATERIALS AND METHODS

Description of the study Area

Lake Hardibo drainage basin (Figure 1) is situated in the northeast escarpment of South Wollo at the western margin of the Afar triangle, 427 km northeast of Addis Ababa, Ethiopia. It is located between 11.19° and 11.28° N, and 39.73° and 39.79° E, found within the Awash River Basin.

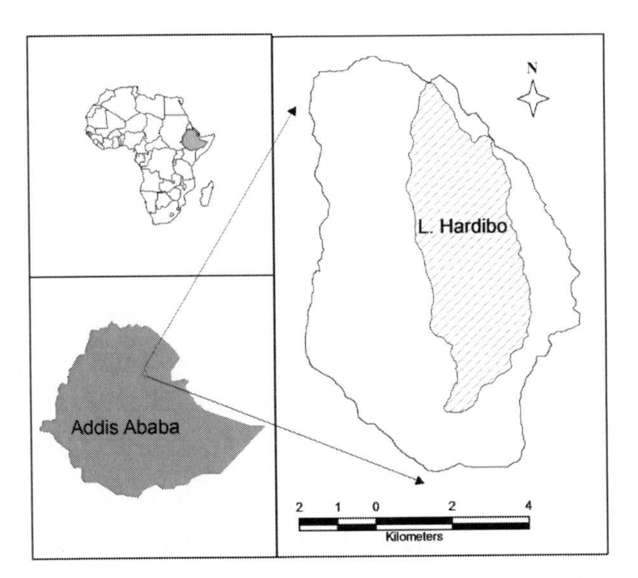

Figure 1. Location map of Lake Hardibo drainage basin.

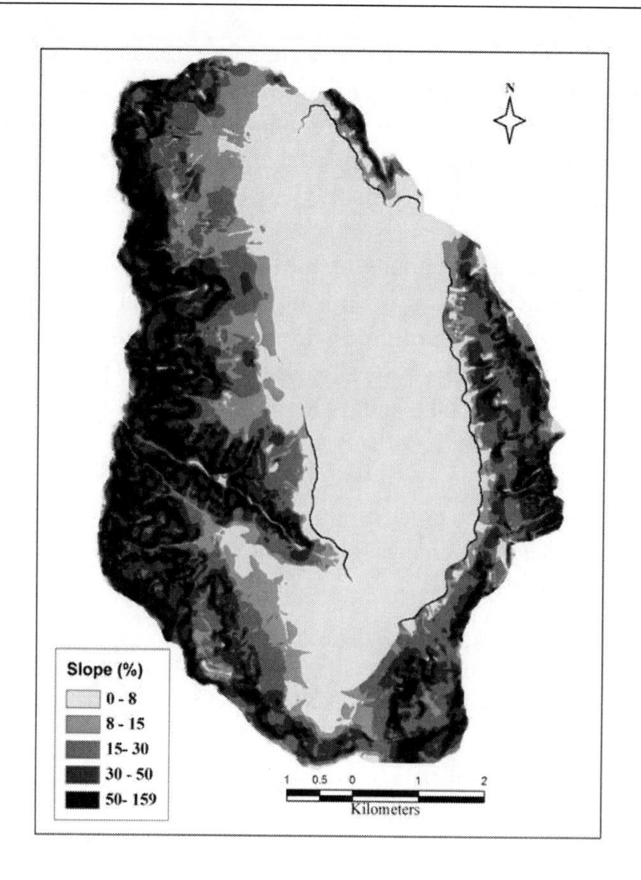

Figure 2. Slope map of Lake Hardibo drainage basin.

Materials

The study utilized two sets of panchromatic aerial photographs acquired in December 1957 and 1986, and another set of 5 m resolution 2007 Spot 5 satellite image of true color and four sheets of 1:50,000 scale topographic maps. The remotely sensed geospatial data were obtained from Ethiopian Mapping Agency, where the satellite image was radiometrically and geometrically corrected, but the aerial photographs were raw.

Methods

Eighteen 23 cm x 23 cm black and white hardcopy photographic images were scanned with high metric quality photogrammetric scanner at 800 dots per inch, dpi (31.75 microns) to produce digital images used for photogrammetric processing as well as visual photo-interpretation activities.

A true natural color of 5-meter resolution Spot satellite image and orthorectified photo images were visually interpreted to produce the final maps. A graphical representation of the sequence of the methods is presented in Figure 3. A change analysis was carried out for two periods, from 1957 to 1986 and from 1986 to 2007.

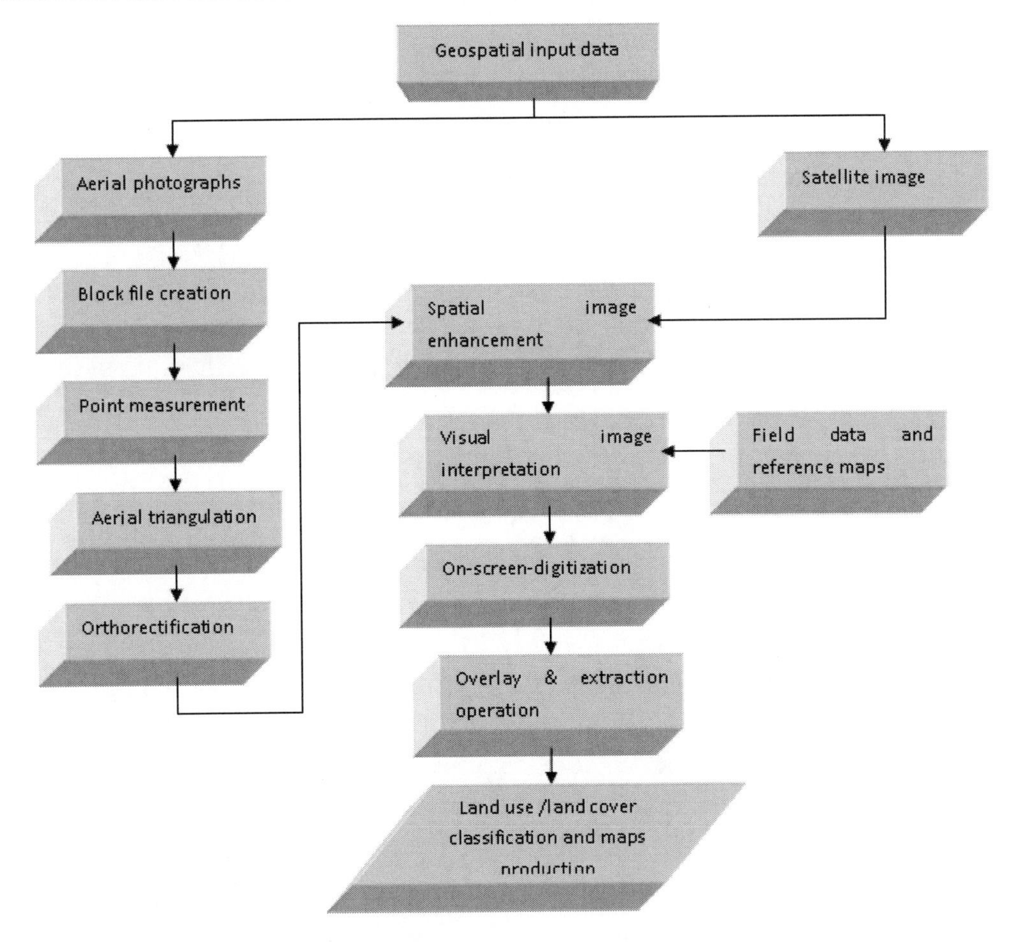

Figure 3. Flowchart of the methodology and analyses.

Tools Used for the Analyses

In this study, principles of Remote Sensing (RS) and Geographical Information System (GIS) techniques were adopted to rectify the raw data and to collect reliable geospatial information (Konecny 1994). Leica Photogrammetric Suite (LPS) version 9.2 Project Manager Software was used to process the raw digital aerial photographs ranging from block formation to the production of orthorectified images. Orthorectification in LPS Project Manager generates planimetrically true orthoimages in which the displacement of objects due to frame camera orientation, terrain relief, and other errors associated with raw images acquisition and processing has been removed (ISPRS 2009). ERDAS Imagine 9.2 and Arc GIS 9.3 were utilized to enhance, visualize and produce maps.

Positional Accuracy Assessment

Location accuracy is normally measured as a mean square error (RMSE). RMSE provides an estimate of the spread of a series of measurement around their assumed (true) values (Figure 4). It is therefore, commonly used to assess the quality of absolute orientation of photogrammetric models and the spatial referencing of satellite imagery (Huisman and de By 2009).

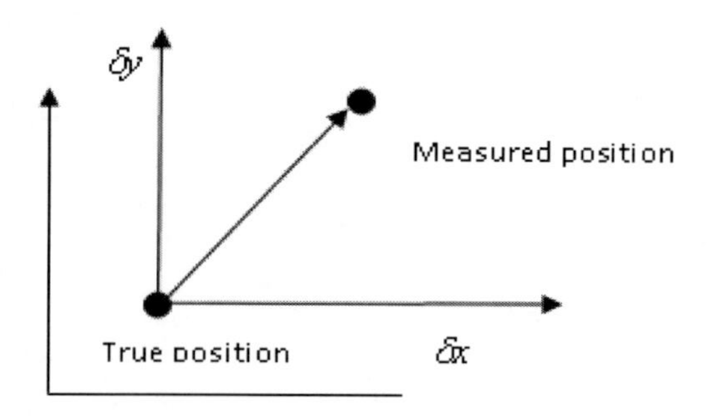

Figure 4. Graphical representation of positional error.

The systematic error in *x* direction and $\delta\bar{y}$ in *y* direction are average deviations from their true value:

$$\delta\bar{x} = \frac{1}{n}\sum_{i=1}^{n}\delta x_i \ , \tag{1}$$

$$m_x = \sqrt{\frac{1}{n}\sum_{i=1}^{n}\delta x_{i=1}^2} \ , \tag{2}$$

The total RMSE was obtained with the formula:

$$m_{total} = \sqrt{m_x^2 + m_y^2} \tag{3}$$

Which, by Pythagorean rule, is the length of the averaged (root mean squared) error.

Block File Creation

Prior to performing photogrammetric tasks within LPS Project Manager, a block was created by entering all necessary information. Camera models, Ethiopian reference coordinate system, camera specific information (Omega, Phi and Kappa), photographic direction, average flying height, and a pair of image principal point were input.

The numbers of fiducials located on an image are four and their photo-coordinate values were measured. After the multiple images were added to the block file and creations of hierarchical pyramid layers were performed, both the internal and external information associated with the cameras were defined. Interior orientations of the pixel coordinate positions of the fiducial marks on each image within a block were predicted in LPS. The RMSE, which measure the overall association between the original fiducial marks' coordinates and their predicted image coordinates, were not larger than 0.5 pixels for all images.

Two exterior orientation parameters: positional elements, X_o, Y_o, and Z_o, defining perspective center of the photo coordinates and rotational elements, Omega (ω), Phi (φ), and Kappa (κ) were input. Finally, a regular block of 4 x 2 photographs and 5 x 2 photographs were formed into 2 overlapping strips for 1957 and 1986 events, respectively covering the study area.

Point Measurements

Point Measurements (Figure 5) were carried out by measuring Ground Control Points (GCP) and Tie Points (TP), which were later used to process aerial triangulation and the subsequent orthorectification steps that established a relationship between the images in the project, the camera, and the ground.

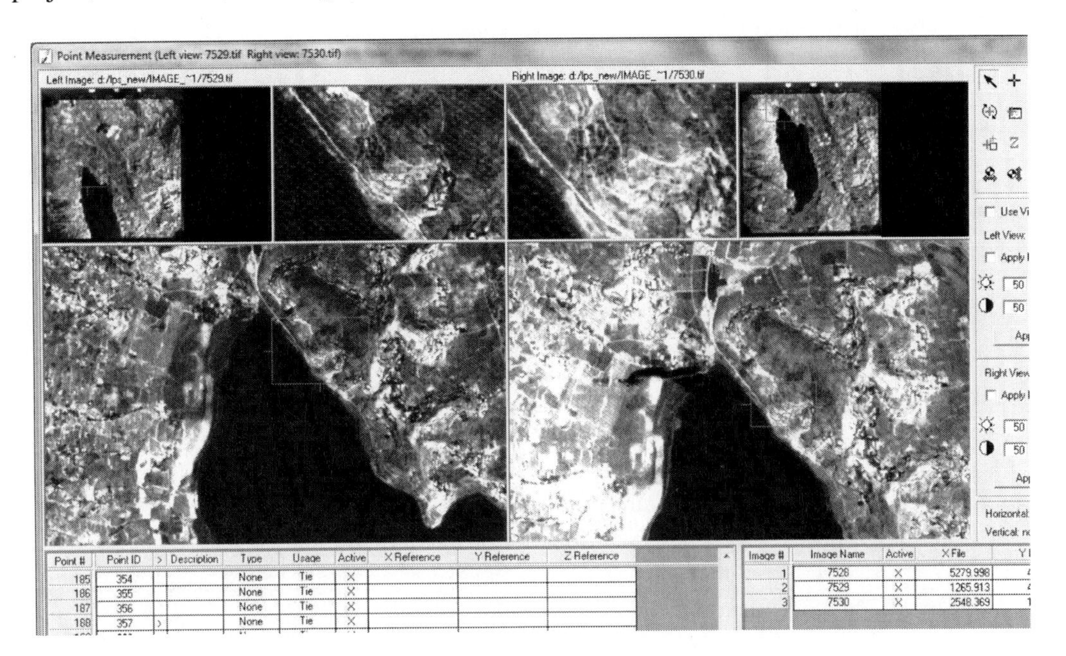

Figure 5. Measurement of GCP and tie points on image ID 7442 and 7443. (LPS dialog window).

Assigning Statistical Weights

As recommended by LPS manual (ERDAS 2009), for small scale mapping projects (greater than 1:40,000), a convergence value of 0.01 m was used as a threshold to determine the level and extent of processing during the iterative aerial triangulation procedure. Statistical properties (standard deviation and statistical weights) were associated with the input data (GCPs, image coordinate points, and exterior orientations) to reflect the quality of the data. Same weighted values were assigned to GCPs to specify a uniform statistical weight

(Table 1) because all the GCPs were surveyed using 3 to 5 m accuracy Garmin hand held GPS. Same weighted values were also assigned to exterior orientation. The strength of the geometric network of exterior orientation comprising the block of imagery did not also vary because all exterior positional parameters were estimated from perspective centers of 1:50,000 topographic maps. Similarly, rotational angles were assigned by comparing the axis of the photo-coordinate system (as defined in the data strip) with the orientation of the images. The extent of the deviation of the parameter values initially measured and finally estimated (by aerial triangulation technique) were controlled by the specified standard deviation values assigned as recommended by the manual , but with more relaxed values. No statistical weights were given to interior orientation parameters because interior orientation parameters (focal length, and principal points in the X and Y direction) were obtained from camera calibration report.

Table 1. Statistical quality parameters

Statistical parameter	Parameter										
	Point measurements					Exterior Orientation					
	GCP, m			Image/photo Coordinates, pixel		Position, m			Rotation, °		
	X	Y	Z	X	Y	X_o	Y_o	Z_o	Omega	Phi	Kappa
Standard deviation	5.00	5.00	5.00	0.95	0.95	5.00	5.00	5.00	0.10	0.10	0.10
Statistical weight	Same weighted value					Same weighted value					

Aerial Triangulation

Aerial triangulation was performed using bundle block adjustment procedure to define a relationship between the images contained within a block, the camera model that obtained the images, and the ground to create accurate imagery and geographic information concerning the study area. With Aerial triangulation exterior orientation parameters (X_o, Y_o, Z_o, and Omega, Phi and Kappa), and coordinates of the tie points (XYZ) were estimated.

Orthorectification

Orthorectified images are planimetrically true images that represent ground objects in their true, real-world X and Y positions (Bakker et al. 2009; ERDAS 2009). For these reasons, orthorectification was executed in LPS Project Manager for triangulated image sequentially using a 30 m resolution ASTER Global Terrain Model (ASTGTM) as a source of Digital Terrain Model (DTM) using bilinear interpolation resampling method to consider the effects of topographic relief displacement. The orthorectified imagery then was used as the ideal reference images backdrop necessary for the creation and analysis of vector data within GIS 9.3 working environments.

Spatial Image Enhancement

Filtering operation was carried out for both orthorectified photos and Spot satellite image in ERDAS Imagine 9.2. A 3x3 Low Pass and 3x3 Edge Enhance smoothing Kernels were used for noise reduction and edge enhancement and optimal images were produced for interpretation.

Visual Image Interpretation

The general methodology employed in this study involved preliminary interpretation, fieldwork, land use and land cover class identification, final interpretation, and map preparation. Prior to delineation of mapping units, an interpretation legend was constructed based on interpretation elements.

Table 2. Description of land use and land cover classes

Land use and land cover classes	Description
Farmlands/settlements	This class is a spatially mixed mapping unit, where farmlands and rural settlements are spatially mixed in a fuzzy manner that needs generalization of information due to interpretation problem of a class to assign a single polygon. Farmlands exist inside the homesteads of dispersed settlements. The class refers to areas where the natural vegetation has been removed or modified and replaced by other types of vegetative cover of anthropogenic origin. The cover is artificial and requires human activities. All vegetation that has been cultivated or planted and intended for harvest was included in this class (examples are annual crop, rural family housing, and qat (chat).
Grasslands	Areas with permanent grass cover or pasture, used for open grazing and meadow. This class also contains grassed peripheral areas of Lake Hardibo that are transitional between terrestrial and aquatic systems during dropping and rising of lake level within a year.
Bush lands	This class consists of an intricate mixture of both small trees and bushes; herbaceous plants and grasses are there in between.
Shrub lands/degraded lands	This class is also a spatial mixed mapping code, where shrub lands and degraded areas are spatially mixed which cannot be distinguished as separate geographic entity and required generalization of information. The Areas were covered with short shrubs and thorny bushes without any defined main stem. Within the short shrubs, 20 to 50 % of degraded areas are present that have open herbaceous plants or grasses in wet season and bare in dry season. Included are bare rock areas mainly with classical gullies.
Forestlands	Areas where the vegetative cover is in balance with biotic and abiotic forces of its biotope. The areas were covered with trees of well-defined stem carrying more or less defined crown, forming nearly a closed canopy (60 to 70 %). This category also incorporates smaller woody plantation, mixed with short bushes, shrubs and open areas. *Juniperus procera* and *Olea Africana* are the main tree types.
Orchard	This class refers to very small fringe areas of Lake Hardibo planted with perennial crops such as fruit trees and sugar cane.
Waterbody	This class refers to open waterbody of the Lake Hardibo that is naturally covered with water.

To have sufficient information about the historical land use and land cover situations in three observation periods, intensive field visits were made from Jan 2009 to Feb 2011 over the entire study area supported by local knowledge. Seven thematic classes were identified and described (Table 2).

The classifiers used in this study were presence of vegetation, edaphic condition and artificiality of cover, as described by Land Covers Classification System (Agrawal et al. 2003; Di Gregorio 2005; Bajracharya 2010).

Based on the correlation between the collected field data and the preliminary interpretation, the entire area was delineated from the orthorectified, spatially enhanced, geometrically and radiometrically corrected satellite image and aerial photographs using on-screen digitization in ArcGIS 9.3. The digitized vector data have undergone also editing, overlay analysis, and topological checks. Finally, a cartographically finished series of static maps, which depict processes of change of land use and land cover, were produced.

In many instances, land cover and land use are treated as equivalent although they have different definitions and are not directly interchangeable. Land use is the present or past use of the land and is defined as “man’s activities on land which are directly related to the land”. Where as, land cover is the type of vegetation or man-made material covering the land surface. These terms are often combined to provide the land use/land cover (LU/LC) classification system (De Barry 2004).

RESULTS AND DISCUSSION

Land Use/Land Cover Extent between 1957, 1986 and 2007

The area of Lake Hardibo drainage basin is 5214.57 ha, including the surface area of the lake. Based on the temporal land use/land cover data produced, the spatial coverage of each mapping unit was shown (Table 3; Figure 6; Figure 7). Farmlands/settlements class was the second in area extent exceeded by waterbody in 1957, but later expanded and contained large area in 1986 and 2007 followed by waterbody, covering a size of 28.10, 48.03, and 39.53% out of the total area in 1957, 1986, and 2007, respectively.

Table 3. Spatial extents of land use and land cover classes from 1957 to 2007

	1957		1986		2007	
Land use/ land cover classes	ha	%	ha	%	ha	%
Farmlands/settlements	1465.17	28.10	2504.42	48.03	2061.27	39.53
Grasslands	155.79	2.99	73.79	1.42	84.95	1.63
Bushlands	1230.31	23.59	309.83	5.94	599.04	11.49
Shrublands/degraded lands	449.04	8.61	693.58	13.30	677.62	12.99
Forestlands	221.98	4.26	1.28	0.02	200.56	3.85
Orchard	0.00	0.00	0.00	0.00	8.31	0.16
Waterbody	1692.27	32.45	1631.67	31.29	1582.82	30.35
Total	5214.56	100.00	5214.57	100.00	5214.57	100.00

The reason behind is that the livelihoods of the population living in the watershed is dependent on subsistence agriculture, practicing cultivation of land for annual crops since thousands of years. The lake waterbody constituted a portion of 32.45%, 31.29%, and 30.35% in 1957, 1986, and 2007, respectively.

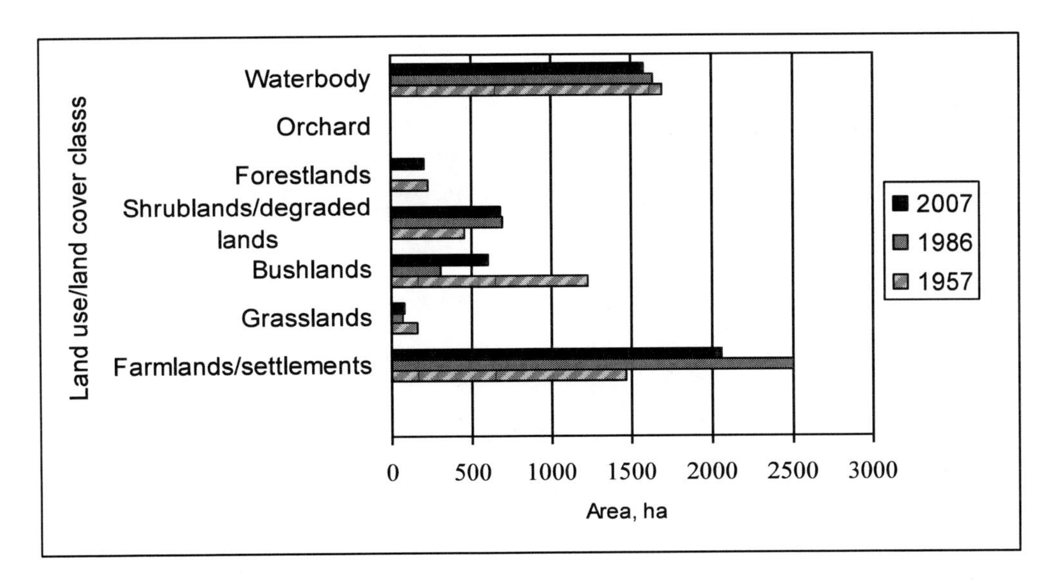

Figure 6. Bar graph showing land use/land covers in 1957, 1986 and 2007.

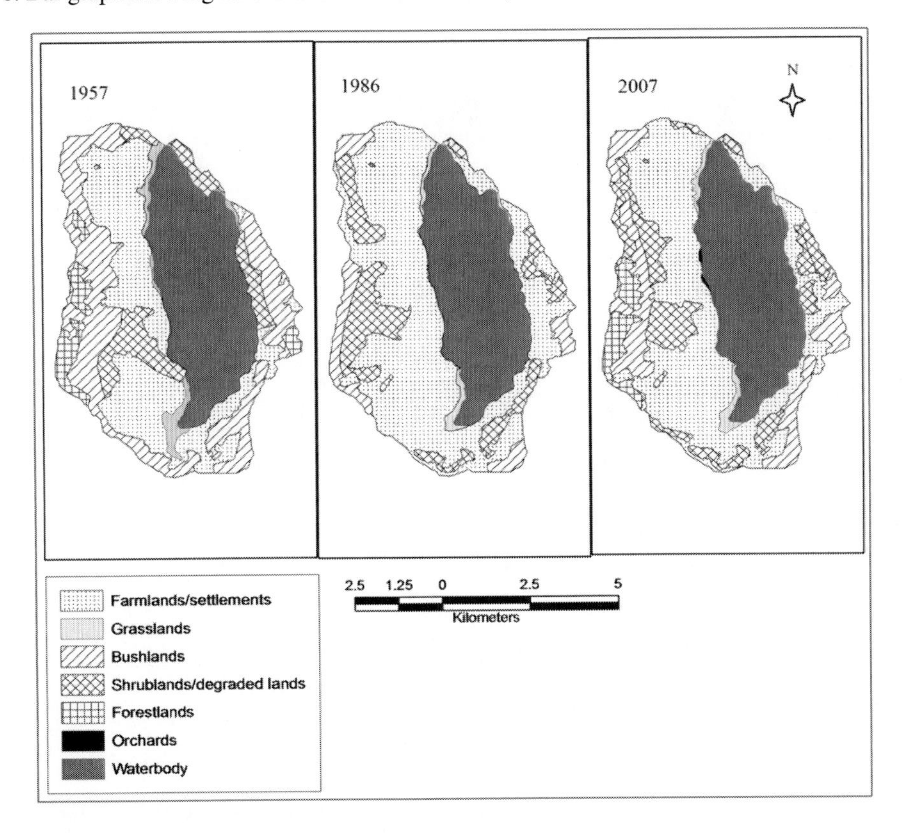

Figure 7. Land use/cover maps in 1957, 1986 and 2007.

Bushlands were the third largest land cover class in 1957 where as shrublands/degraded lands have replaced this rank in 1986 and 2007, showing trends of deforestation through time. Bushlands and shrublands/degraded lands are found on steeper sides of mountains and hills, which are not suitable for cultivation. Forestlands constitute a small proportion in the watershed as was the case in northern parts of the country (Pearson 1992; Darbyshire et al. 2003), comprising of 4.26%, 0.02%, and 3.85% in 1957, 1986, and 2007, respectively. The ruminant trees such as *Juniperus procera* and *Olea africana* species are largely restricted to sacred places and mountaintops. A newly introduced land use type is orchard plantation (0.16%) which was emerged in 2007 in the western fringe of Lake Hardibo. As compared to sufficient water resource potential of Lake Hardibo for small-scale water abstraction and the presence of suitable lands for irrigation around the lake, the present practice for orchards is insignificant.

Land use/land cover changes: 1957 to 1986

Significant land use/ cover changes took place over the 50 years' analysis period in lake Hardibo closed drainage basin (Table 4). Farmlands/settlements have shown the largest increment (+70.93%) with respect to spatial coverage (1039.24 ha) in 29 years period (1957 to 1986) as was the case in most of land use/land cover studies in northeast Ethiopia (Tekle and Hedlund 2000; Zeleke and Hurni 2001; Bewket 2002).

The incremental change in farmlands/settlements was pursued by shrublands /degraded lands (+54.46%, representing 244.55 ha) from 1957 to 1986 at the expense of bushlands (-61.06%), grasslands (-52.63%), forestlands (-99.42%) and waterbody (-3.58%) where expansion of cultivated land encroached the permanent vegetation cover in the steep mountains due to population growth that enforced them to exploit marginal areas.

Table 4. Land use/ cover changes over the half century

Land use/ cover class	Change in land use/land cover classes						Rate of change (1957 to 2007)
	1957 to 1986		1986 to 2007		1957 to 2007		
	ha	%	ha	%	ha	%	ha/y
Farmlands/ settlements	1039.24	70.93	-443.15	-17.69	596.09	40.68	11.92
Grasslands	-82.00	-52.63	11.16	15.12	-70.84	-45.47	-1.42
Bushlands	-920.48	-74.82	289.21	93.34	-631.28	-51.31	-12.63
Shrublands/degraded lands	244.55	54.46	-15.96	-2.30	228.58	50.91	4.57
Forestlands	-220.70	-99.42	199.28	15589.40	-21.42	-9.65	-0.43
Orchards	0.00	0.00	8.31	-	8.31	0.00	0.17
Waterbody	-60.60	-3.58	-48.85	-2.99	-109.45	-6.47	-2.19

The lake surface area shrunk by -3.58% (60.60 ha) in this period, being a typical indicator for misuse of the drainage basin. Others such as decrease of forestland, grassland, bushland

and increase of farmlands and degraded lands are also indicators of unsustainable land management practiced in the basin.

Land use/land cover changes: 1986 to 2007

On the contrary to changes occurred from 1957 to 1986, the expansion of spatial coverage of farmlands/settlements was decelerated (Table 4) by 17.69% from 1986 to 2007. The reduction in expansion of farmlands/settlements in the later 21-year period could be ascribed to the Ethiopian Government's villagization and settlement program practiced in 1980s, which forced demolishing of villages constructed on steeper mountains where farmers were accustomed to cultivate the homesteads and distant lands for annual crops regardless of the slope steepness of the land. Reconstruction of villages on relatively low-lying flat areas and resettlement of some farmers to other regions of the country lessened the burden of marginal mountain areas and gave chance to regenerate natural vegetation.

The rate of transformation from grasslands, bushlands, and forestlands to farmlands/settlement, shrub lands/degraded forms of land use/cover was also drastically declined from -52.63 to +15.12, -74.82 to +93.94, and -99.42 to +15589.40%, respectively from the periods of 1957 to 1986 and 1986 to 2007. Rather, these classes have gained land from farmland/settlements, shrublands/degraded lands and waterbody. The rate of change of waterbody continued to dwindle though the magnitude of changes from 1986 to 2007 (-2.99%) were low as compared to the situation from 1957 to 1986 (-3.58%).

Land Use/Land Cover Changes: 1957 To 2007

The overall land use/land cover dynamics intensified during the last half-century (1957 to 2007). Farmlands/settlements have expanded by 596.09 ha (40.68%) at the rate of 11.92 ha/y taking place on every slope classes. Production of crops amongst households was not sufficient to fulfill food requirement amid spreading out of farmlands. This reveals that the productivity of cropland on steeper slopes where new farmlands expanded were obviously poor. Shrublands/degraded lands were the other increased class unit expanded by 228.58 ha (50.91%) at the rate of 4.57 ha/y by gaining extra lands from bush, grass and forestlands.

Grasslands, which are located at foot slopes in poor drainage areas, were diminished to -70.84 ha (-45.47%) at a rate of -1.42 ha/y. The expansion of farmlands along with settlements and reduction of grasslands is another measure for assessing the competition between crop and livestock production for land resources. The trend showed that crop production was more competitive for land than livestock. During fieldwork, watershed inhabitants told that livestock possession was reduced and at the same time, overgrazing on limited grasslands was practiced due to overall human population increment that caused dwindling of grazing lands.

Bushlands, which are relatively woody perennial plants, have been reduced as much as -631.28 ha (-51.31%) at the rate of -12.63 ha/y, which are mainly transformed to farmlands/settlements and shrublands/degraded lands. Despite the fact that forestlands constituted very small spatial extent from the very beginning in 1957 (4.26%), the watershed underwent deforestation, declined to the extent of -21.42 ha (-9.65%) at the rate of -0.43 ha/y.

The lake surface area was diminishing by -109.45ha (-6.47%) at the rate of -2.19 ha/y owing to the general disturbance of the permanent vegetation in addition to other factors, including may be climate change, and rainfall variability, where human activities induced immediate responses.

Implications of Land Use/Land Cover Changes

Demographic factors play an important role in the evolution and transformation of past, present as well as future land use/land cover, as land use is the product of the interaction of man and land. In the absence of food self-sufficiency, off-farm employment, productive land use policy, prolific land tenure system, sound and sustainable land management system, population pressure impairs the ecological integrity of fauna and flora resulting in multitude of problems, including severe land degradation, soil erosion and reduction of surface and sub-surface water inflows within a watershed.

Demographic data related to population size was not reported in this study because the data complied by Ethiopian Central Statistical Authority were based on administrative structure which did not conform to the boundary of this drainage basin. However, to provide an insight into the magnitude and rate at which the population has been increasing, the size of population of South Wollo is the highest (15.75%) among all Zones of the Amhara Regional State, growing at the rate of 2.8% (CSA 1997). According to South Wollo Zone Department of Agriculture report (1999 unpubl. data), the district, where 90% of this study watershed is located, is the most populated area of all districts in the zone with a population density of 285 persons/ km^2.

The growing demands of the population for food crop and fuel wood have possibly led to land use and land cover changes occurred in the last fifty years, where farmlands, shrub and degraded lands got the momentum to continuously rise with a proximate cause of land degradation and soil erosion in the lake's mountainous drainage basin. The lake intimately connects to its drainage basin, so what happens in the drainage basin is happening in the lake. Denudation of vegetation coupled with inappropriate farming practices have caused soil erosion in the basin and transportation of sediment-laden runoff into the lake (Figure 8).

The hydrology of inflowing streams was also altered due to land use/land cover changes. Lake Hardibo, situated 7 km southeast from Lake Hayq and 231 m higher than Lake Hayq, used to overflow to Lake Hayq through a stream locally named as Ankerkah sometime before the last five decades (personal communication with prominent elders of the watershed inhabitants 2009). That means the two lakes were located within the same closed drainage basin before fifty years ago connected hydrologically by Ankerkah stream. This indicates the declining of surface and subsurface flows into the lakes.

Currently, all tributary streams are ephemeral. The contribution of ground water inflows into the lake water probably also reduced due to devoid of vegetation and poor land management activities which hinder infiltration of surface runoff, a source of lateral and ground water flow, by affecting physical properties of the soil. Therefore, the shrinkage of the lake by -2.19% can be attributed at least partly by land use/land covers changes over the study period. Other negative consequences are certainly loss of land productivity and reduction in the size of livestock due to lack of feed resources resulted in decline of income and food from animals and their products.

Figure 8. Sediment laden runoff into Lake Hardibo (Photo by Hassen Yesuf, 2010).

CONCLUSION

Transformation of land use/land cover is a multifaceted process characterized by varying intensity of factors over time, ranging from land cover class type, geographical situation, local and regional policy environments and socioeconomic conditions. Detailed analyses of land use/land cover class; transformation and implication in the study area have been identified.

Space-borne remote sensing data were used in Lake Hardibo closed drainage basin to monitor spatio-temporal land use/land cover changes using photogrammetric and GIS techniques. Aerial photographs of 1957 and 1986 and satellite image of 2007 have been processed using a combination of Leica Photogrammetric Suite (LPS 9.2) and Geographical Information System (GIS 9.3). The study has produced baseline and derived thematic maps on land use/cover of the lake basin and assessed their effects on the spacio-temporal extent of the lake.

Lake Hardibo has seen significant land use changes over the last 50 years as a consequence of changes in land use/land cover in the drainage basin. Farmland/settlements were the most expanded category of land use/land cover, with a net increment of 40.68%, causing accelerated erosion and soil loss in the fragile landscape of the basin. Devoid of vegetation is taking its toll. Bushlands, grasslands and forest cover have been reduced to a net decline of 51.31%, 45.47%, and 9.65%, respectively. Reduced vegetation cover hinders infiltration of surface runoff, a source of steam flow, which diminishes cumulative water inflows into the lake, moreover denudation of vegetation affect the physical properties of the soil and have been the cause of soil erosion in the basin and transportation of sediment-laden runoff into the lake in the form of silt accumulation. As a result, the surface area of the lake was diminished by 110 ha over the last five decades.

The scientific information generated can help better understand the physical dynamics of the lake and its basin to guide stakeholders and lake managers at the local level for best management practices and help design effective management strategies and policies at higher levels.

REFERENCES

Akililu, A.; Leostroosnijder, and Graaft, J. (2007). Long-term dynamics in land resource use and driving forces in Beressa watershed, highlands of Ethiopia. *Journal of Environmental Management* 83(2007):448-459.

Agrawal S.; Joshi, P. K.; Yogita Shukla; Roy, P. S. 2003. Spot vegetation multi temporal data for classifying vegetation in south central Asia. *Indian Institute of Remote Sensing. VOL. 84, NO. 11, 10.*

Bajracharya, B.; Uddin, K.; Chettri, N.; Shrestha, B.; and Siddiqui, A. (2010). Understanding Land Cover Change Using a Harmonized Classification System in the Himalaya. *Mountain Research and Development*, 30(2):143-156.

Bakker, W.; Feringa, W.; Gieske, A.; Gorte, B.; Grabmaier, K.; Hecker, C.; Horn, J.; Huurneman, G.; Janssen, L.; Kerle, N.; Meer, F.; Parodi, G; Pohl, C.; Reeves, C.; Ruitdnbeek, F.; Schetselaar, E.; Tempfli, K.; Weir, M.; Westinga, E.; Woldai, T. (2009). *Geometric operations*. In *Principles of Remote Sensing*. K. Tempfli, N. Kerle, G. Huurneman, . L. Janssen, eds.; ITC: Enschede, The Netherlands, 4th ed., Series 2, pp. 219-251.

Bewket, W. (2002). Land dynamics since the 1950s in Chemoga watershed, Blue Nile Basin, Ethiopia. *Mountain research and development.* Vol. 22 (3): 283-289.

Bormann, H.; Breuer, L.; Graff, T.; and Huisman, A. (2007). Analyzing the effects of soil property changes associated with land use changes on the simulated water balance: A comparison of three hydrological catchment models for scenario analysis. *ScienceDirect, Ecological Modeling.* Vol. 209: 29-40.

CSA (Central Statistical Authority). (1997). *Statistical Abstract.* Addis Ababa, Federal Democratic Republic of Ethiopia.

Darbyshire, I.; Lamb, H.; and Umer, M. (2003). Forest clearance and regrowth in Northern Ethiopia during the last 3000 years. *The Holecene.* 13 (4): 537-546.

De Barry, PA. (2004). Watersheds: *Processes, Assessment and Management.* John Wiley and Sons, Inc., Hoboken, NJ.

Demlie, M.; Ayenew, T.; and Wohnlich, S. (2007). Comprehensive hydrological & hydrogeological study of topographically closed lakes in highland Ethiopia: The case of Hayq & Hardibo. *Journal of Hydrology.* (339): 145-158.

Di Gregorio, A. (2005). *Land Cover Classification System (LCCS),* Version 2: Classification Concepts and User Manual. FAO Environment and Natural Resources Service Series, No 8. Rome, Italy: Food and Agriculture Organization.

ERDAS. (2009). *Earth Resources Data Analysis System (ERDAS),* LPS Project Manager User's Guide. 5051 Peachtree Corners Circle Suite 100 Norcross, GA 30092-2500, USA.

FAO. 1998. Food and Agriculture Organization of the United Nations. Soil and terrain database for Northeastern Africa, CDROM. FAO, Rome.

Ghaffar, A. (2005). Monitoring land use change: use of remote sensing and GIS. *Pakistan Geographical Review,* Vol. 60 (1): 19-26.

Huisman, O.; and de By, R.A. (2009). *Principles of Geographic Systems. ITC Educational Textbook series, Fourth edition.* ITC, Enschede, The Netherlands.

Hurni, H. (1987). Options of steep land conservation in subsistence agriculture. In: Soil and water conservation on steep lands. San. Jav. Puerto Rico.

Hurni, H.; Tato, K.; and Zeleke, G. (2005). The Implications of changes in population, land use, and land management for surface runoff in the upper Nile Basin of Ethiopia. *Mountain Research and Development.* 25(2): 147-154.

ISPRS (International Society for Photogrammetry and Remote Sensing). (2009). Available from URL: http://www.isprs.org/society.html (accessed Oct 25, 2009).

Konecny, G. (1994). "*New Trends in Technology, and their Application: Photogrammetric and Remote Sensing From Analog to Digital.*" Paper presented at the Thirteenth United Nations Regional Cartographic Conference for Asia and the Pacific, Beijing, China.

Messerli, M.; Hurni, H.; Wolde-Semayat, M.; Tedla, S.; Ives, J.D; and Wolde-Mariam, M. (1988). African Mountains qnd Highlands: Introduction and Resolutions. *Mountain Research and Development,* Vol. 8, Nos. 2/3, 1988, pp. 93-100.

Pearson, CJ. (1992). *Cereal-based systems of the highlands of Northeast Africa.* In Pearson, C. J. (ed). Ecosystems of the world 18: field crops systems, Amesterdam: *Elsevier,* pp 277-289.

Riebsame, W.E.; Parton, W.J.; Galvin, K.A.; Burke, I.C.; Bohren, L.; Young, R.; and Knop, E. (1994). Integrated modeling of land use and cover change. Global impacts of land cover change. American *Institute of Biological Sciences. Bioscience.* 44 (5): 350-356.

Tekle, K.; and Hedlund, L. (2000). Land cover changes between 1958 & 1986 in Kalu District, south Wollo, Ethiopia. *Mountain Research and Development.* 20 (1): 42-51.

Zeleke, G.; and Hurni, H. (2001). Implication of land use/land cover dynamics for mountain resource degradation in the northwestern Ethiopian highlands. *Mountain Research and Development.* Vol. 21(20):184-191.

In: Water Conservation
Editor: Monzur A. Imteaz

ISBN: 978-1-62808-993-6

Chapter 10

CHALLENGES IN THE TRANSITION TOWARD ADAPTIVE WATER GOVERNANCE

Kofi Akamani*

Department of Forestry, Southern Illinois University, Carbondale, IL, US

ABSTRACT

Recognition of the complexity and unpredictability inherent in the management of water resources has triggered the search for alternative institutional mechanisms for water governance that can sustain desired social and ecological values in times of gradual and rapid change. Adaptive governance of water resources has emerged as a promising approach to building the capacity for water resource governance systems to learn and adapt to change across scales. However, evidence of successful transitions toward adaptive governance of water resources is limited. This chapter identifies and discusses various metaphysical, epistemological, institutional, and planning challenges that constrain the transition toward adaptive governance of water resources. Using these themes, analysis of institutional reforms in the Cache River Watershed in southern Illinois shows that in spite of efforts since the 1970s to establish new organizational structures for managing water resources in the watershed, the new resource management regime has retained key features of the conventional resource management paradigm, such as failure to recognize complexity, neglect of local knowledge, lack of community participation, and a narrow focus of resource management goals. Identification of the challenges in adaptive water governance using multiple levels of analysis from the philosophical to the practical level offers a comprehensive and deeper understanding of the sources of rigidity in conventional resource management and promises to inform better policies on how the transition process can be navigated.

* Corresponding author: Kofi Akamani. Tel: +1618-453-7464; Fax +161-8453-7475; E-mail address: k.akamani@ siu.edu.

INTRODUCTION

Growing recognition of the complexity and unpredictability of climate change impacts on the management of water resources has triggered the search for alternative management mechanisms that can sustain desired social and ecological values in times of gradual and rapid change (Pahl-Wostl 2007, Engle et al. 2011). Scientists and policy makers have devoted significant attention to the concept of adaptive management as a framework for guiding the sustainable management of water resources in unpredictable futures. Williams (2011: 1371) defines adaptive management as "learning through management, with management adjustments as understanding improves." As opposed to past resource management approaches that emphasized the efficient utilization of resources through mechanisms of prediction and control, adaptive management recognizes the uncertainties entailed in water resources management and employs a structured decision-making and implementation process that promotes learning (Huitema et al. 2009, Allen et al. 2011, Williams 2011). In spite of its popularity, challenges in the implementation of adaptive management initiatives have largely been attributed to the neglect of the social and institutional context that is needed to facilitate this learning-oriented approach (McLain and Lee 1996, Benson and Garmestani 2011, Cosens and Williams 2012).

Adaptive governance has emerged as a promising institutional framework within which adaptive management could be successfully implemented. Drawing from insights from research on common pool resources and social ecological systems, adaptive governance provides a framework for linking individuals, organizations and other actors across various levels of the jurisdictional scale in an on-going process of policy formulation and implementation in response to gradual and episodic forces of change (Folke et al. 2005, Gunderson and Light 2006, Olsson et al. 2006). Adaptive governance, when considered together with adaptive management, provides mechanisms for enhancing the resilience and robustness of social-ecological systems by building their capacity for adaptability and transformability (Walker et al. 2004, Folke et al. 2010, Walker 2012). It has been argued that the transition from conventional resource management approaches towards adaptive governance mechanisms that can support ecosystem scale adaptive management is critical to sustainable development (Folke et al. 2010, Westley et al. 2011).

However, evidence of such transitions toward adaptive governance regimes is limited (Olsson et al. 2008, Gelcich et al. 2010). A key theoretical explanation for the tendency for incremental change in existing resource management institutions rather than transformative changes is path-dependency (Young 2006, Berkes 2007a, Gelcich et al. 2010). According to Heinmiller (2009: 131), "Path dependency suggests that investments and adaptations in early resource management institutions can make it difficult for actors to abandon these institutions, thereby influencing and shaping subsequent collective action efforts." Thus, institutional actors do not operate on a blank slate but rather, their choices and actions are constrained by a range of contextual factors, such as history, culture and politics (Berkes 2007a). These seemingly stable patterns of interaction may, however, occasionally be punctuated by trigger events that offer windows of opportunity for institutional transformation (Young 2006, Gelcich et al. 2010). The need to understand the process of transition, as well as the challenges in the process is becoming a major research priority (Pahl-Wostl et al. 2007, Sendzimir et al. 2008, Gelcich et al. 2010).

This chapter identifies and discusses a range of metaphysical, epistemological, institutional, and planning challenges that contribute to the rigidity of the conventional water resources management paradigm and constrain the transition toward adaptive water governance. Knowledge of these barriers in the adaptation and transformation process could potentially enhance our understanding of the transition process and inform better policies aimed at building the capacity for learning and adapting to change (Pahl-Wostl et al. 2009, Moser and Ekstrom 2010). The next section of the paper clarifies the conceptual underpinnings of the paper. Next, the major challenges are identified and discussed. The following section analyzes institutional reforms in the Cache River Watershed. Brief concluding comments are then presented in the final section.

Resilience, Adaptability, and Transformability in Social-Ecological Systems

From the ecological perspective, Walker et al. (2004) define resilience as the ability of a system to absorb disturbance and reorganize while undergoing change without compromising its key attributes, such as function, structure, identify, and feedbacks. In coupled human-environment systems, resilience is used to refer to the amount of disturbance a system can absorb while remaining within a given state, the capacity of the system for self-organization, and the ability of the system to build the capacity for learning and adaptation (Folke et al. 2002, Folke 2006). Adaptability and transformability are critical components of resilience that refer to different degrees of change in social-ecological systems (Gunderson et al. 2006).

Adaptability is defined by Walker et al. (2004) as the collective ability of actors in a given system to manage resilience. It entails the capacity of social-ecological systems to learn and apply the knowledge in responding to drivers of change while remaining within the existing stability domain (Folke et al. 2010). Transformability, on the other hand, is defined as "the capacity to create untried beginnings from which to evolve a new way of living when existing ecological, economic, or social structures become untenable" (Walker et al. 2004: 5). Transformations involve long term, deeper changes (Moser and Ekstrom 2010) that result in the crossing of a broad range of thresholds in social-ecological systems, such as meanings and perceptions, social interaction patterns, and institutional arrangements beyond specific resource arenas (Folke et al. 2010). Crossing thresholds may result in irreversible changes in the functioning of the system (Walker et al. 2009). Although transformation may occur either as a planned deliberate action or may result from crossing system thresholds inadvertently, transformational change occurring inadvertently as a result of the cascading effects of changes in other thresholds is likely to lead to undesirable outcomes (Kinzig et al. 2006, Nelson et al. 2007).

Adaptive governance provides the institutional mechanism for building the adaptive and transformative capacity of social-ecological systems for initiating planned adaptive and transformative changes. However, as has been noted earlier, the path-dependent effects of past institutions may constrain the transition to adaptive governance, resulting in a lock-in of the existing system until it is too late (Walker et al. 2004). These sources of rigidity are identified and discussed in the next section.

CHALLENGES

Metaphysical

Efforts to promote planned adaptation and transformation through adaptive governance and adaptive management require embracing resilience thinking (Folke et al. 2010). Resilience thinking is founded upon a complex systems view of social and ecological systems which is fundamentally different from the reductionist assumptions that influence conventional resource management policies. Much of twentieth century resource management policies were influenced by the balance of nature paradigm (Berkes 2007b) that assumed that undisturbed nature existed in an equilibrium state and that nature progressed linearly towards its original equilibrium state following a disturbance event. Management practices built on these assumptions aimed at the constant and predictable flow of products and economic benefits. The dominant application of technical and engineering solutions to water resources problems is a manifestation of the belief in prediction and control (Sendzimir et al. 2008, Pahl-Wostl et al. 2009). Although initially successful, policies based on these assumptions have resulted in unintended ecological consequences (Holling 2000, 2012).

The assumption that humans can be considered distinct from nature is now considered flawed (Redman 2004, Folke et al. 2010). Similarly, the notion that nature is predictable and controllable has been rejected (Folke et al. 2002, Berkes 2007b). The concept of social-ecological systems recognizes that social and ecological systems are inextricably interconnected and interact with each other in a co-evolving manner across spatial and temporal scales (Folke 2007). Panarchy theory helps explain the dynamics of change and stability resulting from the interactions among various components of complex social-ecological systems across scales (Holling 2001, Holling 2012). The focus of adaptive governance on learning and building the capacity for adaptation and transformation is a response to the inherent unpredictability and incomplete knowledge about complex social-ecological systems, such as water management regimes (Pahl-Wostl 2007, Huitema et al. 2009). However, failure of decision-makers and other policy participants to understand complexity often translates into a lack of political support for adaptive approaches to resource management (Walters 2007), as well as the implementation of policies that seek to entrench the conventional resource management paradigm (Sendzimir et al. 2008).

Epistemological

Conventional resources management approaches, characterized by disciplinary isolation and the dominance of experts trained in western science, offer a limited understanding of the complexity of social-ecological systems (Armitage et al. 2009). Complex social-ecological systems may be conceptualized as a series of inter-dependent, semi-autonomous layers, with each layer having its own unique emergent properties (Berkes 2004). As a consequence, no single level of observation offers an adequate representation of the whole (Berkes 2004, Young 2006). Knowledge about such systems tends to be dispersed among actors (Olsson et al. 2007) and dependent upon the scale at which observations are made (Cash et al. 2006).

The integration of knowledge from multiple perspectives across multiple scales, therefore, offers better prospects for a more accurate understanding of such systems.

The adaptive governance of social-ecological systems requires the integration of diverse forms of knowledge (Gunderson and Light 2006). The awareness that expert knowledge offers only a partial and incomplete account of the whole offers justification for the increasing attention being given to traditional and local ecological knowledge (Olsson and Folke 2001). Similarly, various forms of disciplinary collaboration, such as multidisciplinary research, interdisciplinary research, and transdisciplinary research are being pursued to generate integrated knowledge about social-ecological systems (Eigenbrode et al. 2007, Miller et al. 2008). However, this pursuit of knowledge integration raises several critical issues. Irreconcilable differences in the epistemological foundations of western science and traditional knowledge, coupled with the lack of trust and equal recognition of traditional knowledge hampers integration efforts (Young 2006, Berkes 2009). Similarly, efforts aimed at disciplinary collaboration are constrained by a range of individual, disciplinary and institutional challenges (Morse et al. 2007).

Institutional

North (1990) defines institutions as the formal and informal constraints that shape social interactions. Organizations, on the other hand, are groups of people pursuing a common interest. Conventional resource management institutions emphasize the dominant use of centralized institutions as panaceas to conservation challenges. The widespread failure of these centralized institutions, as manifested in various forms of environmental degradation and conflicts, has been attributed to the mismatch between institutions and the scales at which problems occur, limited flexibility in responding to the dynamics of social-ecological systems, capacity constraints, non-participation and so forth (Wondolleck and Yaffee 2000, Karkkainen 2005, Folke et al. 2007). Adaptive governance requires institutional mechanisms that are able to connect actors across multiple scales in an on-going process of learning and adapting to gradual and abrupt forces of change. Dietz et al. (2003) have proposed analytic deliberation, nesting and institutional variety as the three strategies for meeting the institutional requirements of adaptive governance.

Analytic deliberation refers to decision-making procedures designed to combine scientific analysis with public deliberation. It involves a structured dialogue among participants of the policy process, such as scientists, agencies, resource users, and other interested stakeholders throughout the various stages of policy formulation and implementation (Dietz et al. 2003, Balint et al. 2011). Such an approach is likely to yield more accurate information and better decisions because of the integration of diverse forms of ideas and knowledge from policy participants. Through the integration of divergent knowledge systems and values, analytic deliberation is also an effective mechanism for conflict management (Balint et al. 2011). Importantly, analytic deliberation can promote social learning and the shaping of values to facilitate collective responses to change (Akamani and Wilson 2011). However, successful implementation of analytic deliberation procedures will require effective skills in conflict management, including facilitation, negotiation and mediation. Power differentials among participants may also serve as a constraint if not addressed. Other challenges include the lack of political will of agencies and individuals in power to support the process, limited opportunities for effective participation, lack of trust among participants, as well as resource constraints (Wondolleck and Yaffee 2000).

Nesting of institutions across multiple scales is an appropriate response to the realization that social-ecological systems are scale-dependent, and that attention to issues at any single level will be inadequate in dealing with the complexity resulting from the cross-scale interactions (Akamani and Wilson 2011). Similar to nesting, the need for institutional variety is a response to the recognition that no single type of institution, whether community-based institutions, market-based institutions or centralized institutions can serve as a panacea against the multiple drivers of change that social-ecological systems need to respond to (Berkes 2007a). Polycentric institutions provide an appropriate framework to meet these institutional requirements for adaptive governance. Polycentric institutions are semi-autonomous institutions, nested across multiple levels with multiple centers of decision-making and implementation authority (McGinnis 1999, Ostrom 1999, Mostert 2012). A polycentric governance regime may comprise different types of institutions with overlapping functions across multiple scales from the local to the global (Anderson and Ostrom 2008). The nesting of diverse institutions, such as those found in an adaptive co-management framework provides several benefits, including enhancing rule compliance, facilitation of learning and experimentation, and building of capacity for adaptation and transformation (Dietz et al. 2003, Olsson et al. 2004, Armitage et al. 2009). However, such complex institutional frameworks may be challenged by the costs and difficulties involved in coordinating activities across the multiple jurisdictions (Huitema et al. 2009).

Planning

Conventional resource management has traditionally focused on the optimization of benefits from selected components of ecosystems (Cosens 2012). In the field of water resources management for instance, river basin management has traditionally aimed at managing for single purposes, such as navigation and hydropower (Raadgever et al. 2008). The practice of addressing a narrow range of problems in isolation from one another neglects the long term social and ecological consequences of such management practices (Pahl-Wostl et al. 2007).

The scope of adaptive governance covers the broader social context within which ecosystem management occurs (Folke et al. 2005). Adaptive governance departs from a narrow focus and embraces the interconnectedness of social, ecological, economic and political issues entailed in resource management (Gunderson and Light 2006). As such, it seeks to "provide a set of composite policies or solutions that address and integrate these different problem domains (Gunderson and Light 2006: 330). Thus, the transition toward adaptive governance of water resources calls for embracing the holistic perspective of Integrated Water Resources Management (IWRM) in addition to the learning orientation of adaptive management. IWRM aims at coordinating and integrating the management of land and water resources across multiple jurisdictions and temporal scales so as to meet social and economic objectives without compromising the ecological integrity of river ecosystems (Pahl-Wostl et al. 2007, Medema et al. 2008).

Previous efforts to implement IWRM have, however, been hampered by conceptual ambiguities as well as institutional challenges (Medema et al. 2008). Biswas (2004) has forcefully argued that the lack of clarity in the meaning of IWRM, coupled with the challenges posed by bureaucratic politics and limited capacity to manage the complexity of organizational challenges limit the applicability of the concept. More recently, Engle et al. (2011) have argued that the pursuit of IWRM could potentially reduce the capacity of water resource management regimes for learning and adapting to change.

CASE STUDY: THE CACHE RIVER WATERSHED

The Cache River Watershed (Figure 1) covers an area of 1944 square kilometers in the southern part of Illinois (Adams et al. 2005). Due to its unique location at the intersection of four physiographic regions, the watershed contains different types of ecosystems that support a diversity of species (Duran et al. 2004). Since the arrival of European settlers in the watershed in the early 1800s a number of social, economic, technological and institutional factors have resulted in various patterns of land use change in the watershed (Duran et al. 2004). Beside changes in forest cover, the hydrologic features of the watershed have also been significantly modified by various engineering projects. For instance, the construction of the Post Creek Cutoff in 1915 which allowed for the draining of thousands of acres of land for logging and agricultural purposes, basically divided the watershed into two: the Lower Cache River and the Upper Cache River. This was followed by numerous other drainage projects in the watershed, ranging from dredging and channelization to the construction of levees and reservoirs (Cache River Watershed Resource Planning Committee 1995). By the 1970s, the watershed had largely been logged and drained, resulting in various ecological consequences, such as flooding, sedimentation and decline in waterfowl migration patterns to the watershed (Bridges et al. 2007). These problems triggered various institutional reforms.

Since the 1970s, a number of actors across various levels of scale have influenced transitions in the institutional mechanisms and decision processes in the watershed. The Illinois Department of Natural Resources (IDNR) first purchased land in the watershed for conservation purposes in 1970 (Adams et al. 2005). Subsequent years saw increased purchases of land for conservation by various organizations, such as The Nature Conservancy (TNC). The TNC played an influential role in bringing in the US Fish and Wildlife Service (USFWS) which subsequently established the Cypress Creek National Wildlife Refuge in the Cache River Watershed in 1990 (Adams et al. 2005). In 1991, these organizations formally came together to form the Cache River Joint Venture Partnership (JVP). Membership of the JVP currently comprises: IDNR, TNC, USFWS, Ducks Unlimited; and the USDA Natural Resource Conservation Service (NRCS). The activities of the JVP involve land acquisition and restoration with the ultimate goal of reconnecting the Upper and Lower Cache River Watersheds (Davenport et al. 2010).

Between 1993 and 1995 a planning process was implemented to prepare a resource management plan for the watershed (Adams et al. 2005). The process was led by TNC and the NRCS with funding provided by the Illinois Environmental Protection Agency. The procedure for preparing the plan followed a nine-step planning process used by the NRCS (Cache River Watershed Resource Planning Committee 1995).

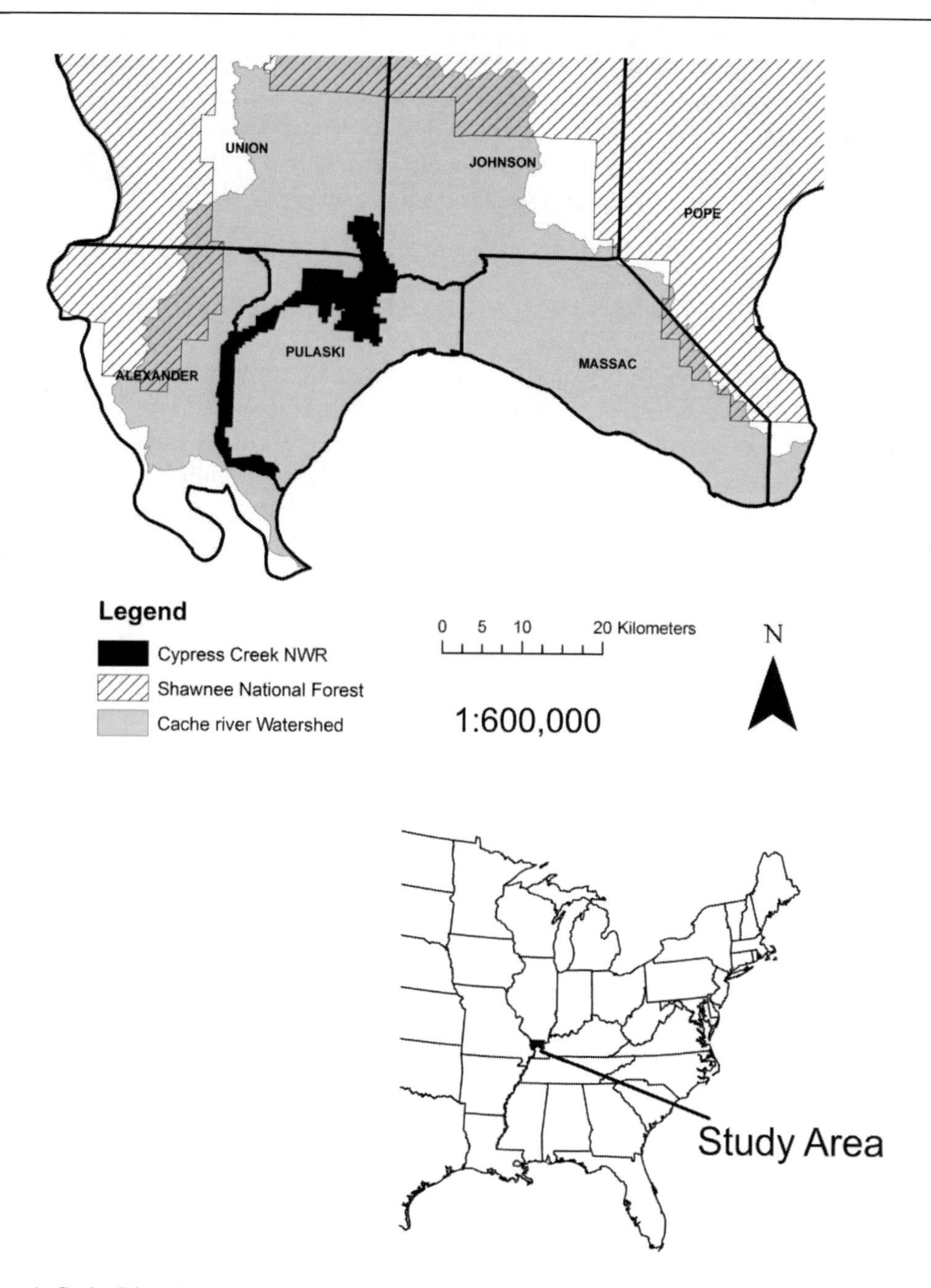

Figure 1. Cache River Watershed.

The process started with the formation of a 25-member Cache River Watershed Resource Planning Committee that comprised mostly of commercial farmers who were selected to represent the five counties in the watershed (Cache River Watershed Resource Planning Committee 2005). Another committee that played a critical role in the preparation of the plan was a 15-member Technical Committee, headed by TNC and the NRCS and comprising expert representatives of 11 organizations, including Southern Illinois University and most of the members of the JVP. In 1995, the Cache River Watershed Resource Plan was completed.

The plan contained a list of nine priority resource concerns and recommendations for addressing them. They are: erosion; open dumping; private property rights; water quality; government farm conservation programs; the Post Creek Cutoff; open flow on the Cache River; dissemination of information; and the impacts of wildlife on farming (Cache River Watershed Resource Planning Committee 1995). Following the completion of the planning process in 1995, the Planning Committee was dissolved. A non-profit organization, Friends of the Cache River Watershed, was then formed to coordinate the implementation of the plan (Adams et al. 2005). Friends of the Cache River Watershed identifies itself as a non-profit citizens' group that works with land owners and members of the JVP with the goal of protecting and restoring 60,000 acres of land along the Cache River (Friends of the Cache River Watershed 2013).

In all, changes that have occurred in the Cache River Watershed since the 1970s appear transformative at first glance. However, a closer look at the current situation in the watershed reveals that legacies of the old paradigm of water resources management still persist in the current management regime. First, the goals and underlying assumptions for managing the Cache River Watershed do not reflect current thinking about the complexity of water management regimes and the need to manage water resources using an integrated and adaptive approach. During the planning process, TNC and NRCS as heads of the Technical Committee framed the planning problem as a resource management problem that did not include the many social and economic problems in the watershed (Lant 2003). The Planning Committee was then charged with the responsibility of coming up with a list of concerns reflecting this resource management problem while the Technical Committee provided input in devising specific recommendations for addressing the issues raised by the Planning Committee (Adams et al. 2005). Although counties in the watershed have some of the highest rates of poverty in the region (Davenport et al. 2010), these issues have largely been ignored and management focus is predominantly on the goal of ecological restoration. Also, there is no explicit recognition of complexity and the need to include learning and adaptation as a priority in the management process. The concluding comments of a strategic plan prepared in 1999 as a supplement to the 1995 Cache River Watershed Resource Plan mentioned the need for adaptive management (Cache River Ecosystem Partnership 1999), although it is yet to manifest in management goals and practice.

Second, the management of the watershed appears to be largely expert-driven with no meaningful efforts to integrate local or traditional ecological knowledge in the management process. As has been noted previously, the goals of the Cache River Watershed Resource Plan as well as the strategies for achieving those goals were largely determined by the Technical Committee that was composed of experts from academic institutions and various government and non-governmental organizations. While most of the commercial farmers who served on the Planning Committee had broader interests in the socio-economic conditions in the watershed, they did not have the power to change the focus of the plan on resource concerns which had been determined by the Technical Committee (Lant 2003, Adams et al. 2005).

Finally, the procedures for decision-making and implementation in the watershed do not meet the institutional requirements of adaptive governance. Membership of the Planning Committee, composed largely of commercial farmers, was not representative of the communities in the watershed and excluded key segments of the local society, such as elected officials and non-landowners (Kraft and Penberthy 2000, Lant 2003). Although public meetings were held by the Planning Committee at the beginning and end of the planning process (Kraft and Penberthy 2000), the resulting plan has been critiqued as lacking legitimacy and not reflective of the needs of communities in the watershed (Adams et al. 2005).

Following the adoption of the plan and the dissolution of the Planning Committee, there has been no sustained mechanism for community participation in subsequent stages of the planning process, such as implementation, monitoring and evaluation. Evaluations of the outcomes of the planning process showed that members of the Technical Committee, such as TNC, IDNR, USFWS and NRCS had successfully used the plan in attracting funding for addressing various resource management challenges such as soil conservation and restoration of the watershed which has resulted in some conservation benefits (Cache River Ecosystem Partnership 1999, Adams et al. 2005). However, community members are largely unaware about the programs being undertaken by members of the JVP and have limited opportunities for participation. The lack of transparency and accountability has given rise to distrust and lack of community support for the restoration activities of the JVP (Davenport et al. 2010).

Conclusion

Lessons from failures in past management approaches, as well as the anticipated impacts of climate change and other drivers of change, point to the need for alternative approaches to water resource management. Adaptive governance offers an innovative institutional framework for connecting individuals and organizations across multiple scales in building the capacity for adaptation and transformation in water resource management regimes. However, research suggests that the path-dependent effects of conventional resource management may constrain the successful transition toward adaptive governance. This chapter has argued that various challenges occurring at the metaphysical, epistemological, institutional, and planning levels interact with one another to constrain the transition from conventional resource management toward adaptive governance regimes. Analysis of the Cache River Watershed illustrates this argument. In spite of the window of opportunity created by ecological crises in the 1970s, the current institutional mechanisms in the watershed appear to have retained key attributes of the old regime. Current focus on ecological restoration while neglecting socio-economic issues and the need for learning reflects a lack of recognition of the complexity of interaction among various components of the watershed. Also, the neglect of local knowledge and community participation in the watershed are reminiscent of the expert-driven, top-down decision-making procedures of conventional resource management.

Folke et al. (2010) have noted that planned transformational change requires an ability to break down the resilience of old regimes and to build the resilience of new regimes. A rich body of literature is accumulating on the role of leadership, enabling policies, incentives, bridging organizations, and the media in the transition process (Folke et al. 2007, Olsson et al. 2007, Olsson et al. 2008, Westley et al. 2011). Such efforts should be channeled towards exploring the role of these elements in creating broad-based awareness about the challenges of complexity in resource management, recognizing windows of opportunity, cultivating interest among policy participants, providing opportunities and building capacities for involvement in the transition process.

ACKNOWLEDGMENTS

I would like to express my gratitude to Dr. Eric Holzmueller for his assistance with the preparation of the maps. Insightful comments from the editor, Dr. Monzur Imteaz, were also instrumental in shaping the focus of the chapter.

REFERENCES

Adams, J., Kraft, S., Ruhl, J. B., Lant, C., Loftus, T., and Duram, L. (2005). Watershed planning: Pseudo-democracy and its alternatives -- the case of the Cache River watershed, Illinois. *Agriculture and Human Values,* 22, 327-338.

Akamani, K. and Wilson, I. P. (2011). Toward the adaptive governance of transboundary water resources. *Conservation Letters,* 4(6), 409-416.

Allen, C. R., Fontaine, J. J., Pope, K. L., and Garmestani, A. S. (2011). Adaptive management for a turbulent future. *Journal of Environmental Management,* 92(1339-1345).

Andersson, K. and Ostrom, E. (2008). Analyzing decentralized resource regimes from a polycentric perspective. *Policy Sciences,* 41(1), 71-93.

Armitage, D., Plummer, R., Berkes, F., Arthur, R. I., Charles, A. T., David-Hunt, I. J., et al. (2009). Adaptive co-management for social-ecological complexity. *Frontiers in Ecology and Environment,* 7(2), 95-102.

Balint, P. J., Stewart, R. E., Desai, A., and Walters, L. C. (2011). *Wicked environmental problems: Managing uncertainty and conflict*. Washington: Island Press.

Benson, M. H. and Garmestani, A. S. (2011). Embracing panarchy, building resilience and integrating adaptive management through a rebirth of the National Environmental Policy Act. *Journal of Environmental Management,* 92, 1420-1427.

Berkes, F. (2004). Rethinking community-based conservation. *Conservation Biology,* 18(3), 621-630.

Berkes, F. (2007a). Community-based conservation in a globalized world. *PNAS,* 104(39), 15188-15193.

Berkes, F. (2007b). Understanding uncertainty and reducing vulnerability: Lessons from resilience thinking. *Natural Hazards,* 41, 283-295.

Berkes, F. (2009). Evolution of co-management: Role of knowledge generation, bridging organizations and social learning. *Journal of Environmental Management,* 90, 1892-1702.

Biswas, A. K. (2004). Integrated water resources management: a reassessment. *Water International,* 29(2), 248-256.

Bridges, C. A., Davenport, M. A., Mangun, J. C., and Carver, A. D. (2007). Cultivating a community-based approach to restoration of the Cache River wetlands in southern Illinois. *Proceedings of the 2007 George Wright Society Conference*, 174-180.

Cache River Ecosystem Partnership (1999). The Cache River Watershed Strategic Resource Plan: A supplement to the 1995 Cache River Watershed Resource Plan.

Cache River Watershed Resource Planning Committee (1995). Resource Plan for the Cache River Watershed: Union, Johnson, Massac, Alexander and Pulaski Counties Illinois.

Cash, D., Adger, W. N., Berkes, F., Garden, P., Lebel, L., Olsson, P., et al. (2006). Scale and cross-scale dynamics: Governance and information in a multi-level world. *Ecology and Society,* 11(2).

Cosens, B. A. (2012). Legitimacy, adaptation, and resilience in ecosystem management. *Ecology and Society,* 18(1).

Cosens, B. A. and Williams, M. K. (2012). Resilience and water governance: Adaptive governance in the Columbia River Basin. *Ecology and Society,* 17(4).

Davenport, M., A., Bridges, C. A., Mangun, J. C., Carver, A. D., Williard, K. W. J., and Jones, E. O. (2010). Building local community commitment to wetlands restoration: A case study of the Cache River Wetlands in southern Illinois, US. *Environmental Management,* 45(4), 711-722.

Dietz, T., Ostrom, E. and Stern, P. (2003). The struggle to govern the commons. *Science,* 302 (5652), 1907-1912.

Duram, L., Bathgate, J. and Ray, C. (2004). A local example of land-use change: Southern Illinois -- 1807, 1938, and 1993. *The Professional Geographer,* 56(1), 127-140.

Eigenbrode, S. D., O'Rourke, M., Wulfhorst, J. D., Althoff, D. M., Goldberg, C. S., Merrill, K., et al. (2007). Employing philosophical dialogue in collaborative science. *Bioscience,* 57(1), 55-64.

Engle, N. L., Johns, O. R., Lemos, M. C., and Nelson, D. R. (2011). Integrated and adaptive management of water resources: Tensions, legacies, and the next best thing. *Ecology and Society,* 16(1).

Folke, C. (2006). Resilience: The emergence of a perspective for social-ecological systems analyses. *Global Environmental Change,* 16, 253-267.

Folke, C. (2007). Social-ecological systems and adaptive governance of the commons. *Ecological Restoration,* 22, 14-15.

Folke, C., Carpenter, S., Elmqvist, T., Gunderson, L., Holling, C. S., and Walker, B. (2002). Resilience and sustainable development: Building adaptive capacity in a world of transformations. *Ambio,* 31(5), 437-440.

Folke, C., Carpenter, S. R., Walker, B., Scheffer, M., Chapin, T., and Rockstrom, J. (2010). Resilience thinking: Integrating resilience, adaptability and transformability. *Ecology and Society,* 15(4).

Folke, C., Hahn, T., Olsson, P., and Norberg, J. (2005). Adaptive governance of social-ecological systems. *Annual Review of Environment and Resources,* 30, 441-473.

Folke, C., Pritchard, L., Berkes, F., Colding, J., and Svedin, U. (2007). The problem of fit between ecosystems and institutions: Ten years later. *Ecology and Society,* 12(1).

Friends of the Cache River Watershed (2013). Http://friendsofcache.org.

Gelcich, S., Hughes, T. P., Olsson, P., Folke, C., Fernandez-Gimenez, E., Foale, S., et al. (2010). Navigating transformations in governance of Chilean marine coastal resources. *PNAS,* 107(39), 16794-16799.

Gunderson, L. and Light, S. S. (2006). Adaptive management and adaptive governance in the everglades ecosystem. *Policy Science,* 39, 323-334.

Gunderson, L. H., Carpenter, S. R., Folke, C., Olsson, P., and Peterson, G. (2006). Water RATs (resilience, adaptability, and transformability) in lake and wetland social-ecological systems. *Ecology and Society,* 11(1).

Heinmiller, T. B. (2009). Path dependency and collective action in common pool governance. *International Journal of the Commons,* 3(1), 131-147.

Holling, C. S. (2000). Theories for sustainable futures. *Ecology and Society,* 4(2).

Holling, C. S. (2012). Response to "panarchy and the law". *Ecology and Society,* 17(4).

Huitema, D., Mostert, E., Egas, W., Moellenkamp, S., Pahl-Wostl, C., and Yalcin, R. (2009). Adaptive water governance: Assessing the institutional prescriptions of adaptive (co-) management from a governance perspective and defining a research agenda. *Ecology and Society,* 14(1).

Karkkainen, B. C. (2005). Transboundary ecosystem governance: Beyond sovereignty? In: C. Bruch, L. Jansky, M. Nakayama and K. A. Salewicz (Eds.), *Public participation in the governance of international freshwater resources.* New York: United Nations University Press.

Kinzig, A. P., Ryan, P., Etiene, M., Allison, H., and Elmqvist, T. (2006). Resilience and regime shifts: Assessing cascading effects. *Ecology and Society,* 11(1).

Kraft, S. and Penberthy, J. (2000). Conservation policy for the future: What lessons have we learned from watershed planning and research. *Journal of Soil and Water Conservation,* 55(3), 327-333.

Lant, C. (2003). Watershed governance in the United States: The challenges ahead. *Water Resources Update,* 126, 21-28.

McGinnis, M. D. (1999). *Polycentric governance and development.* Ann Arbor, Michigan: University of Michigan Press.

McLain, R. J. and Lee, R. G. (1996). Adaptive management: Promises and pitfalls. *Environmental Management,* 20(4), 437-448.

Medema, W., McIntosh, B. S. and Jeffrey, P. J. (2008). From premise to practice: A critical assessment of integrated water resources management and adaptive management approaches in the water sector. *Ecology and Society,* 13(2).

Miller, T. R., Baird, T. D., Littlefield, C. M., Kofinas, G., Chapin III, F. S., and Redman, C. L. (2008). Epistemological pluralism: Reorganizing interdisciplinary research. *Ecology and Society,* 13(2).

Morse, W. C., Nielsen-Pincus, M., Force, J. E., and Wulfhorst, J. D. (2007). Bridges and barriers to developing and conducting interdisciplinary graduate-student team research. *Ecology and Society,* 12(2).

Moser, S. C. and Ekstrom, J. A. (2010). A framework to diagnose barriers to climate change adaptation. *PNAS,* 107(51), 22026-22031.

Mostert, E. (2012). Water management on the Island of IJsselmonde 1000 to 1953: Polycentric governance, adaptation and petrification. *Ecology and Society,* 17(3).

North, D. C. (1990). *Institutions, institutional change and economic performance.* Cambridge, MA: Cambridge University Press.

Olsson, P. and Folke, C. (2001). Local ecological knowledge and institutional dynamics for ecosystem management: A case study of Lake Racken Watershed, Sweden. *Ecosystems,* 4, 85-104.

Olsson, P., Folke, C., Galaz, V., Hahn, T., and Schultz, L. (2007). Enhancing the fit through adaptive co-management: Creating and maintaining bridging functions for matching scales in the Kristianstads Vettenrike Biosphere Reserve, Sweden. *Ecology and Society,* 12(1).

Olsson, P., Folke, C. and Hahn, T. (2004). Social-ecological transformation for ecosystem management: The development of adaptive co-management of a wetland landscape in southern Sweden. *Ecology and Society*, 9(2).

Olsson, P., Folke, C. and Hughes, T. P. (2008). Navigating the transition to ecosystem-based management of the Great Barrier Reef, Australia. *PNAS,* 105(28), 9489-9494.

Olsson, P., Gunderson, L. H., Carpenter, S. R., Ryan, P., Lebel, L., Folke, C., et al. (2006). Shooting the rapids: Navigating transitions to adaptive governance of social-ecological systems. *Ecology and Society,* 11(1).

Ostrom, E. (1999). Coping with tragedies of the commons. *Annual Review of Political Science,* 2, 493-535.

Pahl-Wostl, C. (2007). The implications of complexity for integrated resources management. *Environmental Modeling and Software,* 22, 561-569.

Pahl-Wostl, C., Sendzimir, J. and Jeffrey, P. (2009). Resources management in transition. *Ecology and Society,* 14(1).

Pahl-Wostl, C., Sendzimir, J., Jeffrey, P., Aerts, J., Berkamp, G., and Cross, K. (2007). Managing change toward adaptive water management through social learning. *Ecology and Society,* 12(2).

Raadgever, G. T., Mostert, E., Kranz, N., Interwies, E., and Timmerman, J. G. (2008). Assessing management regimes in transboundary river basins: Do they support adaptive management? *Ecology and Society,* 13(1).

Redman, C. L., Grove, M. J. and Kuby, L. H. (2004). Integrating social science into the Long Term Ecological Research (LTER) Network: Social dimensions of ecological change and ecological dimensions of social change. *Ecosystems,* 7, 161-171.

Sendzimir, J., Magnuszewski, P., Flachner, Z., Balogh, P., Molnar, G., Sarvari, A., et al. (2008). Assessing the resilience of a river management regime: Informal learning in a shadow network in the Tisza River Basin. *Ecology and Society,* 13(1).

Walker, B., Holling, C. S., Carpenter, S., and Kinzig, A. (2004). Resilience, adaptability and transformability in social-ecological systems. *Ecology and Society,* 9(2).

Walker, B. H. (2012). A commentary on "resilience and water governance: adaptive governance in the Columbia River Basin". *Ecology and Society,* 17(4).

Walker, B. H., Abel, N., Anderies, J. M., and Ryan, P. (2009). Resilience, adaptability and transformability in the Goulburn-Broken Catchment, Australia. *Ecology and Society,* 14 (1).

Walters, C. J. (2007). Is adaptive management helping to solve fisheries problems? *Ambio,* 36 (4), 304-307.

Westley, F., Olsson, P., Folke, C., Homer-Dixon, T., Vredenburg, H., Loorback, D., et al. (2011). Tipping toward sustainability: Emerging pathways of transformation. *Ambio,* 40: 762-780.

Williams, B. K. (2011). Passive and active adaptive management: Approaches and an example. *Journal of Environmental Management,* 92, 1371-1378.

Wondolleck, J. M. and Yaffee, S. L. (2000). *Making collaboration work: Lessons from innovation in natural resource management.* Washington, D.C.: Island Press.

Young, O. R. (2006). Vertical interplay among scale-dependent environmental and resource regimes. *Ecology and Society,* 11(1).

INDEX

#

A

B

C

D

G

H

I

J

K

L

M

N

O

P

Q

R

S

T

U

V

W

Y

Z